CATALOG OF HUMAN SOULS
Book 1

HOMO SAPIENS ARE BIO-ROBOTS
Human "Software"

Olga Skorbatyuk and Kate Bazilevsky

HPA Press

ISBN-10: 0996731202
ISBN-13: 978-0-9967312-0-1

Olga Skorbatyuk and Kate Bazilevsky
***Homo Sapiens* Are Bio-Robots: Human "Software"**

Part 3 translated by Kate Bazilevsky
All images are in the public domain unless otherwise noted.

Dedicated to the 40th anniversary of the discovery of the Catalog of Human Population by researcher Andrey Davydov.

Table of Contents

PREFACE

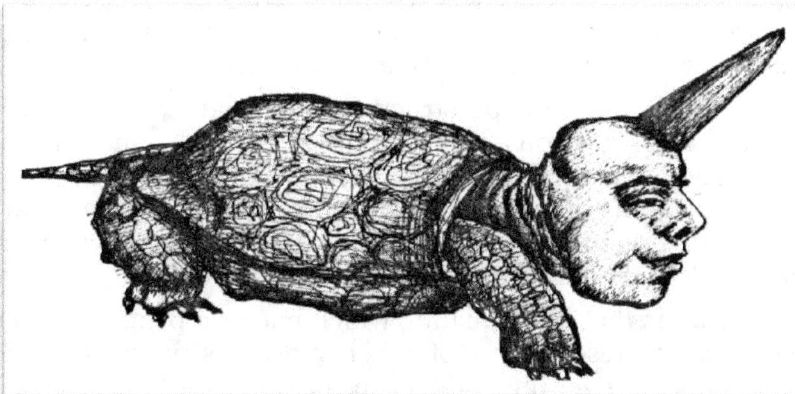

Sketch © 2000 Andrey Davydov

We—authors of the Catalog of Human Souls book series—would like to briefly describe what these books are about right from the beginning because the topic of the Catalog of Human Population is very new for many people.

These books are dedicated to the 40th anniversary from the beginning of research that led to discovery of the Catalog of Human Population. A technology of uncovering of the individual structure of human psyche (or simply put—the soul) was created on the basis of this scientific discovery. The author of scientific discovery of the Catalog of Human Population and this technology is a researcher of ancient books, an expert in Chinese culture Andrey Davydov.

The source of knowledge about the structure of human psyche and the basis for creation of this technology is one of the most ancient and mysterious texts preserved in this civilization—the ancient Chinese monument Shan Hai Jing (translated as the Catalog of Mountains and Seas), which Andrey Davydov managed to decrypt.

After Andrey Davydov created the technology of decryption of Shan Hai Jing (Catalog of Mountains and Seas), it was found that this ancient Chinese monument contains very detailed descriptions of 293 models of human psyche and a lot of other kind of information about the structure of

Homo sapiens. On this basis, Shan Hai Jing was qualified as the Catalog of Human Population.

In Shan Hai Jing (Catalog Mountains and Seas), the biological type "human" is described as a type divided into 293 subtype structures according to the phenological principle. It turned out that each person, belonging to one of these 293 subtypes from birth, has stable characteristics of this subtype; regardless of race, nationality and particularities of parental psychophysiological structures, which are only minor correctors.

Information about the subtype structure is implanted in the form of a program in the unconscious of a person from birth, and this program determines all of his life: his/her personal qualities and character properties, algorithms of life and functioning, hidden motivational spring, abilities, talents, preferences, inclinations, etc. Natural subtype program is that what is called "psyche", "soul."

Natural program is that individuality, which makes a person different from other people as representatives of other subtype structures of biological type *Homo sapiens*. Each individual "speaks his own language," specified by his subtype program, as the language of values, views, convictions, preferences, which are standard and unchanging for all representatives of a subtype.

In the language of science, a natural subtype program of *Homo sapiens* is called Individual Archetypal Pattern; in simple language—an Individual (Subtype) Program, an Individual Program or a Program.

A human program is recorded in the language of natural images. Images or, using the language of science, archetypes of the unconscious sphere of a person is the language of human "software." The concept of "an archetype" was introduced to psychology by Carl Gustav Jung, but as it turned out, archetypes can be not only of the collective unconscious, but also individual. Therefore, to avoid confusion, in popular texts we prefer to call the language of "software" of *Homo sapiens* by the word "image" instead of "archetype."

Programs of each of 293 human subtypes are recorded by different natural images and a different number of images, meaning that they are endemic, are not similar to one another. *Homo sapiens* is a living system, which, as it turned out, exists and functions strictly on the basis of a natural program implanted from birth, and from this it was concluded that a human is a bio-robot at the genetic level.

The conclusion that "*Homo sapiens* is a bio-robot" is confirmed by that learning someone's natural individual (subtype) program from the Catalog of Mountains and Seas as the Catalog of Human Population, it is possible to find out absolutely everything about this person in great detail; about any aspect of his life and activities, including that what he or she carefully conceals.

In addition, *Homo sapiens*, as a biosystem programmed by nature, has modes of self-regulation and regulation (control from the outside). Modes of regulation are a natural inborn mechanism, just like an individual human program. Their discoverer Andrey Davydov named these modes Individual Manipulation Modes (Manipulation Modes for short). This management tool, which can be applied to any person, also was found in the text of the ancient Chinese monument Shan Hai Jing.

It was found that for every person, as a biosystem, there are three manipulation modes: suppressing, balancing and stimulating. Manipulation modes together with an individual (subtype) program are individual structure of psyche of *Homo sapiens*. Programs and manipulation modes of each subtype differ, from one another. For this reason, people differ from each other by internal characteristics, and individual manipulation scenarios are necessary for each person.

Structure Of Psyche Of *Homo Sapiens*

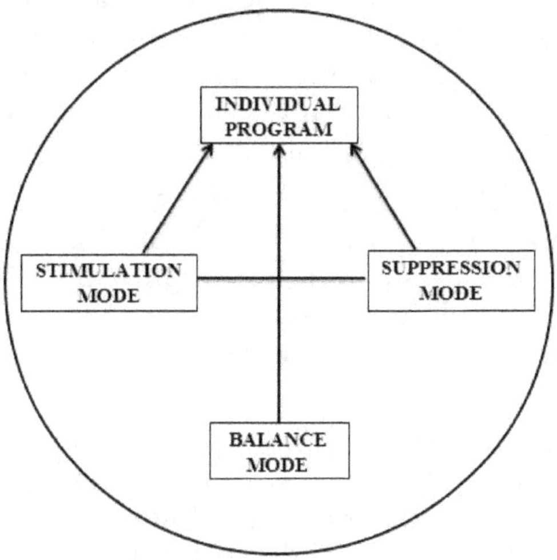

© 2015 Olga Skorbatyuk, Kate Bazilevsky

Individual program, with which a person was born, is the main segment of structure of human psyche (soul). Knowing the natural individual program allows you to find out who he/she (or you personally) is real, without masks, what motives drive this person (or you), and so on.

Suppression mode is a self-control mode, which turns on automatically as a mode of relaxation, pleasure, dream, happy oblivion (a "turn off" button). It is the individual program of people, who belong to a particular subtype.

When a manipulator acts out a personality from the suppression mode, it makes an individual experience pleasure, unconsciously get attached to the manipulator (fall in love, etc.), and on this basis obey him/her and fulfill all of his/her requests/demands.

Balance mode is a self-control mode, which turns on automatically when a person needs to become balanced, get in a harmonious, comfortable state without loss of activeness. It is the individual program of people, who belong to a particular subtype. When a manipulator acts out a personality from the balance mode, it makes an individual experience comfort in communication with him/her, trust him/her at the unconscious level and consider him/her the best friend.

Stimulation mode is a self-control mode, which turns on automatically when a person needs to stimulate himself/herself to some actions, become active. It is the individual program of people, who belong to a particular subtype. When a manipulator transmits properties of the stimulation mode in a certain way, it makes an individual experience strong irritation, up to wild rage, and on this basis he/she does that what a manipulator asks for/demands.

Natural manipulation modes described in Shan Hai Jing (Catalog of Mountains and Seas) as the Catalog of Human Population provide a key to managing any person. With the help of knowledge of human manipulation modes, it is possible to factually change a person's physical and psychological state, behavior, reactions in the desired direction.

No one is able to resist application of his personal manipulation modes because it is a natural mechanism, which is built into human psyche from birth. Therefore, information about manipulation modes of a subtype provides unlimited possibilities for influencing any individual as a representative of this subtype and allows self-control and control of any human as a biosystem.

Non-traditional psychoanalysis, a new direction in scientific psychology, allows identifying an individual's subtype structure (psyche) and manipulation modes on the basis of the Catalog of Human Population. It is called non-traditional because it does not use any of the traditional principles, approaches to the study of human psyche, as well as methods (observation, experiment, testing, biographical method, questioning, conversation) since these methods are not needed in order to obtain any kind of information about a person as a stable and identifiable biosystem.

A non-traditional psychoanalyst knows what a person is like, what problems might concern him or her and why, and how these problems can be effectively solved without any kind of contact with this person. All that a non-traditional psychoanalyst needs to know about the subject being studied is the day, the month, the year of his or her birth and the gender.

Neither race nor nationality or place of residence of the person being studied matter because the Catalog of Human Population contains descriptions of unchanging characteristics of subtypes of *Homo sapiens* of the entire human population.

For colleagues from the scientific environment we are offering our definition of what human psyche and the Catalog of Human Population are: "*The Catalog of Human Population is a description of a human as a type by subtype structures. Subtype structure ("psyche", "soul") is a combination of individual archetypes, recorded at the genetic level (principle). Expressions and interaction of subtype structures in manipulation modes and phenological algorithms are described with adjustments for gender, age and cultural differences. Information is recorded on six factors.*" This definition was developed by Andrey Davydov—the author of discovery and decryption of the Catalog of Human Population.

It is also necessary to tell a bit about the main source of knowledge (knowledge of that *Homo sapiens* are bio-robots, that human psyche (soul) is "software," that "software" of any person can be easily uncovered, and so on), which was obtained in the course of our scientific research. To date, no one knows for sure who gave and preserved this knowledge, but after it was written down it got the title 山海經 Shan Hai Jing (translated from Chinese as Canon/Catalog of Mountains and Seas).

According to Artem I. Kobzev (a Russian historian of Chinese philosophy, Doctor of Philosophical Sciences, professor, author of over eight hundred scientific papers on the history of Chinese philosophy, science and culture), Shan Hai Jing is an anonymous monument, which presumably dates back to the late III century BC-early II century BC. It consists of eighteen juan (scrolls), combined into two sections: Canon/Book of the Mountains (Shan Jing) or Canon of Five Innermost Mountains (Wu Zang Jing) in five juan and Canon of the Seas (Hai Jing) in thirteen juan.

Legend claims that Shan Hai Jing was engraved on sacred vessels by Bo Yi—an assistant to the wise semi-mythical ruler of the ancient times Yu the Great, who lived in the XXIII century BC. Yu the Great entered the throne in 2205 BC. Authors of the Han epoch attribute authorship of the literary monument Shan Hai Jing to Yu the Great and his companion Bo Yi.

According to legend, Yu the Great dealt with a great flood, which fell upon earth and arranged it. The deedful ruler allegedly came to know its mountains, rivers, their spirits, as well as animals and plants. He ordered his assistant to describe everything that was seen. As a result, these recordings together with images of spirits, fantastic animals, birds and plants were engraved on nine ritual vessels-tripods. Later on, these sacred vessels were lost. However, according to historical annals, prior to their strange disappearance, the text of the Catalog of Mountains and Seas, along with amazing images of representatives of flora and fauna, spirits and deities were copied.

This answers the question why some date Shan Hai Jing back to III-II centuries BC, while others back to XXIII century BC—the Catalog of Mountains and Seas as the source of information appeared during the time of Yu the Great and it got the form of a text much later. We tend to agree with those researchers, who date Shan Hai Jing back to XXIII BC since according to Dan Zhu the Catalog of Mountains and Sea was recorded only after its long-term oral existence (this point of view is expressed in his commentaries to the famous monument Spring and Autumn - Chunqiu). Reports that Yu the Great and his assistant Bo Yi created the Catalog of Mountains and Seas exist in, for example, Wang Chong's (27-97 AD) treatise titled Critical Essays: "When Yu and Bo Yi were taming the waters of the flood—Yu was engaged in calming the water, while Bo Yi in recording information about various "things." And, they created the Catalog of Mountains and Seas."

We will add that apparently in mythologies of almost all cultures of the world (those, which to continue to exist, as well as those, which already sank into oblivion) exist facts, which show that the Catalog of Human Population was present in these cultures. This is not difficult to trace by carefully studying the ancient, archaic cultural layers. However, the source itself was preserved only in one culture—the culture of China. This is not surprising since Chinese culture is not only ancient, but also, in spite of everything, the Chinese manage to preserve it and their traditions from the ancient times to the present day (unlike representatives of other cultures).

This was a brief description of the main points, about which you will be able to learn in detail from the Catalog of Human Souls book series.

*****

In essence, the Catalog of Human Souls book series is one book divided into five parts. All these parts are devoted to a single topic—scientific discovery made by researcher Andrey Davydov, the Catalog of Human Population. Division into five parts was done because materials for this book are quite non-uniform in content and style.

The book Catalog of Human Souls is divided into five parts since fundamental research being carried out by Andrey Davydov and our colleagues is interdisciplinary, and the field of research is wide and multifaceted (psychology, gender relations, sociology, political science, linguistics, sinology and so on), and also because we present results of this research not only in the language of science, but also in the language of popular science.

In addition, we (authors of Catalog of Human Souls books), who are presenting research and discoveries of our laboratory's research supervisor Andrey Davydov, found it necessary to include some information about him as well. It seemed to us that the story of discovery of the Catalog of Human Population might be of interest not only to those, who are already familiar with it, but also to those, who are just learning about the Catalog.

Also, in order not to sound groundless, we decided to offer readers to get acquainted with some of materials from the Catalog of Human Population. These materials are provided in the form of brief demonstration versions from descriptions of natural programs of people. We are of the opinion that presentation of any, even the most captivating scientific theory is, as the saying goes, not worth even a straw, if this theory cannot be tested and used on practice.

Demos that we presented certainly are very short and they are by no means complete descriptions of individuals. However, perhaps they will be quite enough for independent testing, which even a non-professional in the field of psychology will be able to do and get a confirmation that the Catalog of Human Souls (a nonscientific title of the Catalog of Human Population) really exists. And, this means that now he or she can use this Catalog in daily life.

We—authors of the Catalog of Human Souls book series—have the honor to not only tell about some of the research done by our laboratory's research supervisor Andrey Davydov and our colleagues, but also to participate in the development of the Catalog of Human Population. However, it must be stated right away that in all sections where we tell about the technology, methodology of non-traditional psychoanalysis—we are only narrators. Technology of decryption of the ancient Chinese monument Shan Hai Jing was created over 20 years ago and its authorship belongs to Andrey Davydov.

*****

The Catalog of Human Souls book series contains the following five books:

Book 1. *Homo Sapiens* Are Bio-Robots. Human "Software" (by Olga Skorbatyuk and Kate Bazilevsky). This book tells about the new method of obtaining information from the unconscious sphere of a human and about the Catalog of Human Population in the form of answers to the most frequently asked questions. In this book, we also presented three hundred selected research topics, which were developed in our laboratory between 1974 and 2014.

Book 2. Hack Anyone's Soul. 100 Demos Of Human Programs From The Catalog Of Human Population (by Olga Skorbatyuk and Kate Bazilevsky). In this book, we are offering one hundred brief demonstration versions taken from descriptions of human programs in the Catalog of Human Population. Of course, these descriptions are only a tiny fraction of information that the Catalog contains about each person. However, we think that this is enough for independent testing and to obtain confirmation of existence of the Catalog of Human Population.

Book 3. Human Manipulation Modes. Either You Are Manipulating Or You Are Being Manipulated (by Olga Skorbatyuk and Kate Bazilevsky). This book is entirely devoted to manipulation modes—the natural toolkit for controlling a system called "a human," which was discovered in the ancient Chinese monument Shan Hai Jing (which turned out to be the Catalog of Human Population). In this book, we explain what individual manipulation modes are and how to use them to manipulate people. After all, with the discovery of the Catalog of Human Population, only two positions in society are left—either you are manipulating or you are being manipulated. And, as proof of that this is really so, in this book we provided the scenario of suppression mode for manipulation of people, who were born on October 12th of leap years or October 13th of common years. Despite that this is only one of four modes of manipulation of those people, who were born on the dates indicated—it will be sufficient for testing.

Book 4. Non-Traditional Psychoanalysis. Selected Scientific Articles And Presentations at Conferences (by Andrey Davydov and Olga Skorbatyuk). This book presents some of the scientific articles and presentations at scientific conferences done by Andrey Davydov, the author of scientific discovery of the Catalog of Human Population. It also includes several scientific articles (which are also chapters of the textbook titled Archetypal Pattern. Fundamentals of Non-Traditional Psychoanalysis.), authors of which are founders of non-traditional psychoanalysis: Andrey Davydov and his colleague—psychologist Olga Skorbatyuk.

Book 5. Shan Hai Jing—A Book Covered With Blood. The Story Of Developers of the Catalog of Human Population (by Kate Bazilevsky). This book uncovers the story of developers of the Catalog of Human Population—Andrey Davydov and Olga Skorbatyuk— that is related to obtaining of political asylum in the United States of America due to many years of persecution by the Federal Security Service of Russian Federation (FSB, former KGB) and attempts to kill them in order to take possession of their research product—the Catalog of Human Population.

*****

From the beginning, as authors of the book series Catalog of Human Souls, we would like to apologize for the lack of a single language style and popular language. Our colleagues from the scientific environment, professional psychologists or those, who are interested in scientific psychology, can satisfy their curiosity about the theory that lies at the basis of technology that we use to identify the individual structure of psyche in Book 4 of this series (Non-Traditional Psychoanalysis. Selected Scientific Articles And Presentations At Conferences.).

The other four books in this series are intended for the widest audience. From our point of view, any person, regardless of education level, must have an opportunity to get acquainted with results of our research. After all, Shan Hai Jing, as the Catalog of Human Population, was left to all humanity. Therefore, in many of our books (despite that they are all devoted exclusively to scientific topics) we try to use the literary language, avoiding frequent use of specific terminology. However, we were unable not to use jargon at all, and for this we beg your pardon. We are not writers, our main work is scientific research, and in books we are sharing with the audience some of the results of this activity, according to the common practice in the scientific community.

Speaking of language that we use to describe the topic of our research... It was noticed that some people are perplexed by that we use the word "soul," as it is not customary to use this word in the scientific community. However, as it was found, human soul, which is the same as psyche and subtype structure—exists. And, it exists regardless of someone's opinions about it.

Soul (psyche), as "software" of *Homo sapiens* is not a psychological, religious or metaphysical value—as it turned out, it is purely a natural value. Therefore, like any natural phenomenon (rain, wind, electricity, gravity), the soul functions regardless of whether people know about it or not, and regardless of what they think about this. People can argue as much as they want whether soul exists or not, and fantasize as much as they want about what it is, but the natural mechanism called "soul" will continue to work. In every person, while he is alive.

Therefore, every person has the right to continue debating whether or not the soul exists and fantasize about what it is, but we prefer to study the human soul as an existing phenomenon (which, by the way, is possible to see, hear, smell and touch, if you have information about individual structures of psyche) and use results of this research on practice.

We consider the question of what words to use to call the natural phenomenon "soul" unimportant when it is possible to study it as a phenomenon instead of talking about it. We use the word "soul" because in any language it accurately reflects the essence of the phenomenon—that foundation, on the basis of which any human being lives, and because of which he is alive; that what is called "closer than the body." And, unlike our colleagues-psychologists, we are not forgetting that "psychology" translated from the ancient Greek means "science about the soul" (from the ancient Greek ψυχή - "soul," λόγος - "teaching").

*****

Now, let's get back to the Catalog of Human Souls books series. Those people, who are already familiar with our research and our other books, usually have questions, to some of which we would like to provide answers in the Catalog of Human Souls book series. For example, people are interested to know who exactly is working on compiling of the Catalog of Human Population, and why it is not possible to find information anywhere, including on the Internet, about all members of our laboratory's staff. Also, people want to know how they can get information from the Catalog of Human Population about themselves or other people.

With regard to the question about developers of the Catalog of Human Population and other staff members of our laboratory, it really is difficult to find reliable information about them and here is why. This happens for two reasons: the first reason is explained using an example with the Internet in the Introduction to Book 1 of this series titled Homo sapiens Are Bio-Robots, and the second reason is confidentiality of all of this kind of information and is related to security. The number people in our laboratory, who work on compiling of the Catalog of Human Population, is not three people, as it might seem. However, unfortunately, we cannot provide any information about who these people are. Reasons for this are detailed in Book 5 of this series (Shan Hai Jing—A Book Covered With Blood. The Story Of Developers Of The Catalog Of Human Population).

Due to persecution because of the main subject of our scientific research—the Catalog of Human Population—by a group of employees of Federal Security Service of Russian Federation (FSB, former KGB), headed by colonel of Foreign Intelligence Service (data for the period 2000-2004) Andrey Dmitrievich Polonchuk, we do not have the right to provide any kind of information about our laboratory's staff. You can get acquainted with the evidence of these prosecutions in Book 5 mentioned above. Originals of documents presented there are stored in the United States Department of Homeland Security, Federal Bureau of Investigation (FBI), etc.

Therefore, if anywhere in public sources, for example on the Internet, you come across a statement that someone is an employee of the Special Scientific Info-Analytical Laboratory—Catalog Of Human Souls or the Human Population Academy, and is engaged in developments related to decryption of Shan Hai Jing and compilation of the Catalog of Human Population—it can be either disinformation or fraud, or both. However, we do have partners and affiliates. Therefore, it is always best to check with us.

Beware that some research works of Andrey Davydov (the author of discovery of the Catalog of Human Population) and the Special Scientific Info-Analytical Laboratory—Catalog Of Human Souls led by him were stolen by a group of employees of the Federal Security Service of Russia (formerly KGB) headed by Andrey Dmitrievich Polonchuk—colonel of SVR FSB of Russian Federation (information about his rank refers to 2000-2004). Therefore, it is possible to find offers on the Internet (in different languages) to purchase information from the Catalog of Human Population, even though this information is in no way related to the Catalog or to natural human "software."

Such offers can be found at the following websites: in Russian – Mountains and Seas of Self-Knowledge http://mountaseas.com, Catalog of Human Population (Catalog of Mountains and Seas) http://vk.com/chp_lab, Find The Answer Within Yourself http://www.facebook.com/groups/mountaseas.2905/, Catalog of Human Population http://vk.com/googlite; in English – Shan Hai Jing Lab/Shan Hai Jing Laboratory http://www.facebook.com/SHJLab/info, http://twitter.com/SHJLab and http://shjlab.wordpress.com; etc.

It is extremely dangerous to buy products offered by fraudsters. Since some time back Andrey Davydov suspected that sooner or later, in one form or another piracy will occur and made some inaccuracies in materials related to Shan Hai Jing; without taking them into account it is impossible to correctly put together the base material.

Perhaps, after realizing this A. D. Polonchuk's group decided to get at least some, at least material benefits from stolen works and this is how a strange, ugly product was born, which has nothing to do with information from the Catalog of Human Population as the decrypted ancient Chinese monument Shan Hai Jing.

The consequence of this is that those, who get information from fraudsters cause irreparable damage to their health; both physical and psychical. Unfortunately, our laboratory already has a significant amount of information about cases when people, who requested information from the Catalog of Human Population from those, who officially have nothing to do with this research, experienced serious damage in various matters and of varying degree of severity. The reason for this is the fact that human psychophysiology functions on the basis of images. Therefore, if a person receives images and decryptions (which essentially are images as well), which are in no way related to his/her personal psychophysiology and begins to use this information on practice, then consequences might be as follows:

- psychical disorders of any kind, which cannot be cured
- somatic disorders of any kind, which cannot be cured

- death, as a result of psychical and physiological disorders

Therefore, please be careful. You have every right to turn to anyone for information from the Catalog, but we issued this warning and we will not be held responsible in any way for possible consequences of your decisions.

You can get information from the Catalog of Human Population about yourself or any other person directly from our laboratory right now. For convenience of our clients, the system of issuance of any kind of information directly from developers of the Catalog of Human Population is made in such a way that they do not need to contact us. Anyone can view prices and pay for an order here - https://www.humanpopulationacademy.org/pricing.

And, for those people, who would like to not just get materials from the Catalog of Human Population sold on our website, but also contact us (for example, with a business proposal, to order of some specific services, to get consultations, to become a student, etc.)—please refer to Human Population Academy's Contacts page at https://www.humanpopulationacademy.org/breakthrough-discovery/contacts/.

Purchasing information from the Catalog through our website or by contacting us are the only ways to get information from the Catalog of Human Population directly from its developer without the risk of coming across fraudsters.

It should be noted that no one ever persecuted our clients; they never were and are not in any danger because those who persecuted and continue to persecute us were never interested in our clients as they do not have information about the technology of decryption of Shan Hai Jing. In addition, we always respect confidentiality of any person, who turns to us for information from the Catalog and we never share information about our clients with anyone.

*****

In conclusion, we would like to state the principled position that we adhere to as authors of books (and, currently there are over three hundred of them)—we do not in any way claim to be gurus; we are more comfortable with the position of students. However, due to the fact that we engaged and continue to engage in research work, we are able to obtain information that can be interesting and useful not only to us. And, we are just legalizing this information for those, who prefer not to believe, but to know. To know how the world works and what is his or her own, personal structure.

Another principal point is that in our books, we do not share our personal opinions, hypotheses. Everything that we state is information from the ancient sources that we are studying and facts obtained during the course of our research activities, which have been tested, as it is customary in scientific practice. We prefer to keep to ourselves our opinions and fantasies, as insignificant in comparison with that what is stated in our books.

Also, on behalf of our colleagues from the laboratory and ourselves, in celebration of the 40th anniversary since the beginning of research of the ancient Chinese monument Shan Hai Jing as the Catalog of Human Population, we would like to express deep gratitude to all of our relatives, friends, acquaintances and colleagues from the scientific environment, as well as to all others, whom we happened to come across in life, for that they failed to prevent us from carrying out this research.

Olga Skorbatyuk and Kate Bazilevsky

Los Angeles, California

June 2015

INTRODUCTION

THERE IS NO SECRET KNOWLEDGE—THERE ARE CHILDREN, WHO DO NOT KNOW HOW TO READ

Sketch © 2000 Andrey Davydov

In this civilization the label "secret," for example, "secret society", "secret knowledge" usually means that there is no knowledge, but it is shameful to admit this. After all, knowledge is recipes. The ancient books are a different matter; books much older than, for example, the Bible, even though Christians consider it ancient for some reason. These books often really do contain knowledge. However, there is no artificial secrecy around them—just open them and read them. Another question is that far from everyone succeeds in reading and hence the conclusion that "this is a great mystery, and it is not given to us to cognize this." To fans of mysticism, esotericism and researchers, who unriddle mysteries using scientific methods, we are offering to look at this from a slightly different angle: secret knowledge does not exist in nature—there are children, who do not know how to read.

*****

The main secret of this civilization is that there is no civilization. After all, civilizations are organized in order to help *Homo sapiens* survive and evolve. And, this strange formation called "civilization," in which we live, during the entire time of its existence has not helped even a single representative of this biological type to survive. People died, and continue to die. They die "without leaving the cradle," if we consider their psychological age. And, what to say about evolution since not a single one of those, who were born in the human form, became a human in this civilization.

This is not our personal opinion. If we rely on criteria of who a human is, which are recorded in the most ancient sources preserved in this civilization, then not a single representative of the biological type *Homo sapiens* currently living on planet Earth is a human. At least, there is no information about this and there is no opportunity to come across a human out on the streets. Figuratively speaking, all around there are only seeds, from which trees could grow, but there are no trees. And, for this we should thank "civilization," in which all of us are "lucky" to live; or to be more precise—not to live, but to survive.

If we put aside the inadequate opinion of *Homo sapiens* about themselves and rely solely on information from these ancient books, then not a single person around is capable of demonstrating human qualities or abilities. Moreover, humanity clearly jumped the gun even with the addition of "sapiens" (from Latin "wise") since by the level of its psychophysiological development, it remains at the stage of *Homo erectus* ("upright man"). It is a pity because such quality as wisdom would significantly help representatives of this biological type, which are stuck in their development, become those, who they were originally "projected," go all the way to the status of a "human."

The status, qualities and abilities of a human, as a being principally different from all those, who we see around us, do not descend from the sky onto *Homo sapiens*, like manna from heaven. And, they never will descend, even if you wait millions of years. This status needs to be attained, and it is necessary to attain it using certain methods. And, for this it is necessary to learn. And, in order to learn knowledge is required. Special knowledge as recipes in the form of "do this and do it this way, and this is what you will get," instead of that what is usually offered in this "civilization" ("go I know not whither and fetch I know not what") and certainly instead of fantasies like "I'm already a human" because no matter for how long you imagine that you are a human—you are not going to become a human. The history of this civilization demonstrated this well.

Through film production, literature, music, etc., there are attempts to motivate humanity, which is driven to the state of settled down animals, to open up their natural potential; probably because even as a resource *Homo sapiens* is becoming not very effective. Even though offered recipes of how to do this are absolutely terrible. For example, consider the movie titled Lucy by Luc Besson (2014).

In the movie Lucy, just like in all other sources of information available in this civilization, microscopic pieces of information are mixed into such an unimaginable salad with disinformation that for an average person it is an impossible task to distinguish one from the other. Although an average person does not even try to do this since he was already told: "this is fiction." However, psychologists have long found that human psyche perceives any fairy tale as reality. By telling fairy tales to children (including adult children), they are being given recipes of how to live and, using computer terms, programs according to which people will unconsciously act get installed in their unconscious.

Either intentionally or due to that psychologists still confuse psyche with intellect, the movie offers the following recipe: a person can become a being with unlimited abilities through activation of his brain to one hundred percent. This certainly sounds tempting to *Homo sapiens*, whose brain is not even ten percent developed. However, intellect is not psyche. Intellect is only a part of the psyche. Psyche consists of six factors: intellectual, physical, nutritional, emotional, sexual, and environmental. If only intellect gets activated, then the same thing will happen with physiology of an individual as with an electric household appliance of one hundred and twenty volts, if it were to get connected without a transformer directly to a high-voltage power line; and without the happy-end "I am everywhere," as it was shown in Lucy.

Also, it was shown that, for example, knowledge about everything, including about his own self, is inside a person, even though knowledge is outside, while inside are only fantasies. And, once again it was demonstrated that the way to open up human potential is through drugs, although not a single drug taker ever became a being with incredible abilities; by using this recipe people usually become corpses, permanent patients of narcologists, crazy or homeless people, living under a bridge. And so on and so forth.

The movie did not in any way mention that the main method to open up human potential is by gaining knowledge, studying and that it takes time for *Homo sapiens* to learn and to grow. As a result, today humanity is a community of Neanderthals—half-animals, who exchanged a stone ax for an “iPhone”. Since *Homo sapiens* continue to be at the level of the Stone Age even when it comes to questions of their own structure, from the standpoint of natural psychical functioning.

It is not surprising that people have problems with reading ancient texts. And, there is proof. One of the very ancient monuments titled the Catalog of Mountains and Seas (Shan Hai Jing), exact dating of which is unknown, but according to some researchers it appeared sometime during XXIII century BC, lay directly "under the nose" of humanity for thousands of years, and no one succeeded in understanding what it is about. Even though quite educated people, for example, the ancient Chinese philosopher Confucius (IV-V centuries BC) were familiar with this ancient monument. As it turned out, it contains that knowledge, which for centuries humanity has been searching for and continues to search for—knowledge about the structure of *Homo sapiens*, the Catalog of Human Population.

*****

When it comes to the Catalog of the human population, for some reason it seems to people that it is something new, even though it is not. Although we do not consider the tale of king Solomon amusing and to us the source of this tale is not authoritative, the well-known expression "there is nothing new under the Sun" (Ecclesiastes 1:9) in the case of the Catalog of Human Population sounds quite reasonable. Since it is very likely that this Catalog was in our civilization from the moment of its emergence, but it simply did not occur to anyone that this ancient monument is a detailed description of structure of psyche of the entire human population.

The principle of that "there is nothing new under the sun" applies to other areas as well. If we put aside childish fantasies of humanity about that people invented many new things and that in this sense they managed to surpass mother-nature itself, it is necessary to state the fact that during the entire history *Homo sapiens* did not create anything that does not already exist in nature. Probably you will ask: what about the “iPhone”? According to information about human structure from the ancient sources, any modern technical equipment is only material realization of unopened psychophysiological potential of a human. Everything that people supposedly invent comes specifically from there, and not from someone's genius head.

However, let’s get back to the Catalog of Human Population. As already stated above, no one thought up or invented it. The ancient Chinese monument Shan Hai Jing, which turned out to be the source of knowledge about the psychical structure of *Homo sapiens*, by all appearances was in this civilization for about forty centuries. And, it was neither kept secret nor hidden—it lay in plain sight. However, many readers of Shan Hai Jing (Catalog of Mountains and Seas), who tried to understand what this text is about, each time made judgments, which were too far from the true state of things.

This ancient monument was taken for a mythological reference, for a geographical atlas, for a bestiary, for ridiculous fantasies of the ancients—

there were most different version, but at the same time they were absolutely identical in that all of them were way off the point. In parallel, humanity searched for answers to questions, which are natural for any individual: "Who am I? What am I like? What was I born for and what am I living for?" While specifically the Catalog of Mountains and Seas contains full and complete answers to all of these questions (and even to those questions, to which humanity has not yet matured).

During many centuries no one of the many researchers of the Catalog of Mountains and Seas were able to understand what this ancient monument is about, until the moment when in the 80s of the XX century this ancient source was decrypted by a Russian researcher, a specialist in Chinese culture Andrey Davydov. Andrey Davydov found out that this ancient monument is nothing other than the description of structure of human psyche—the Catalog of Human Population.

However, if we get back to the question of who created the Catalog of Human Population, then, in essence, all that that this researcher did was that he simply managed to read the ancient text. Andrey Davydov is the author of scientific discovery of the Catalog of Human Population, but he is not the author of the Catalog itself. And, this is big difference. Andrey Davydov and his colleagues are developers of the Catalog not because they are creating it, but because they are creating the decrypted version of the Catalog of Mountains and Seas, which any person can read. Moreover, anyone can use the many recipes, which this source contains, in their daily lives.

While on the topic of decryption of the ancient Chinese monument Shan Hai Jing, it is important to note that Andrey Davydov did not arrange information obtained from it, did not "decorate" it with his own or someone else's comments, opinions, conclusions, additions. Moreover, it was not necessary to make any personal additions in order to clarify something or add to the content of the Catalog of Mountains and Seas since this ancient text originally contains an unimaginably huge amount of information about who a human is and what is the structure of his psyche.

All that had to be done in order for the ancient writings to "begin to speak" and to get an opportunity to use them as an encyclopedia of *Homo sapiens* was to decrypt them. And, this is what was done. However, Andrey Davydov, the author of the decryption technology of Shan Hai Jing, always insisted on that he is only the translator of this ancient text; the translator of the language of archaic images into the language of modern psychological description.

And, the decryption technology of Shan Hai Jing really was developed by Andrey Davydov and he developed it himself without anyone's help. In addition to work with hieroglyphics from the Catalog of Mountains and Seas, this technology also includes work with information from a large number of areas of scientific knowledge; for example, such sciences as phenology, biology, zoology, botany, mineralogy, geomorphology, physical geography, hydrology, linguistics, as well as other natural sciences and the humanities. This technology allowed to return the ancient source of knowledge about a human and his soul (psyche) to this civilization.

The question of whether the technology of decryption of Shan Hai Jing, which Andrey Davydov created in order to obtain descriptions of people and other recipes contained in this ancient source, works or does not work has long been answered, as it has been verified in the course of experiments in scientific institutions of the Russian Federation back in the 90s of the XX century. Since the main rule of decryption of Shan Hai Jing is not to make any kind of addition to information obtained in the course of decryption—descriptions from the ancient source absolutely match characteristics of real people; as it should be, in the case of the Catalog of Human Population.

This fact is observable and can be tested by anyone, who wishes to do so, even a non-professional in the field of psychology. And, we suggest that you do not leave this test for some other time. In the second book of the series Catalog of Human Souls titled Hack Anyone's Soul. 100 Demos Of Human Programs From The Catalog Of Human Population we provided one hundred demo versions of descriptions of natural programs of *Homo sapiens*. Any person can compare these descriptions to people whom he knows well and thus obtain confirmation that the Catalog of Human Population is not fiction.

We do not propose to believe in that the Catalog of Human Population exists and that we have learned to decrypt ancient writings, which turned out to be this Catalog. We are not apologists of another religion that requires faith. We are engaged in science and in science it is necessary to check facts obtained in the course of research and to test on practice.

Descriptions obtained by Andrey Davydov and his colleagues as a result of decryption of the Catalog of Mountains and Seas carry authentic information about real people, who live here and now; about their qualities of personality, life algorithms, functional diapason, philosophy, preferences, abilities and much more. They also contain information about how a particular real person can be managed. This refers to management of behavior and psychophysiological state of a subject of manipulation, but more details on this topic can be found in the third book of this series, which is titled Human Manipulation Modes. Either You Are Manipulating Or You Are Being Manipulated.

Fortunately, the original text of the Catalog of Human Population reached us from ancient times in an entirely preserved form. Perhaps due to that in his time Confucius, like all other previous researchers, failed to understand what Shan Hai Jing describes and categorized this monument as describing evil spirits, demons, with which a good person should have nothing to do.

Thus, he gave this ancient text a writ of protection for many centuries and defended the original text of the Catalog of Mountains and Seas from additions and corrections, which usually hopelessly ruin books.

Left untouched, the text of Shan Hai Jing (Catalog of Mountains and Seas) safely carried knowledge about the structure of human psyche, the Catalog of Human Population in its entirety through the centuries.

*****

From our point of view, questions about what the Catalog of Human Population is and why it is needed sound a bit strange. Especially in the XXI century, when there are catalogs and encyclopedias of anything: plants, animals, cars, real estate, clothing, furniture, etc. However, apparently, the word "catalog" in relation to *Homo sapiens*, his psyche, his soul sounds so unusual to an average person that some explanations are required.

In a nutshell, the Catalog of Human Population is quite a detailed description of qualities of personality and individual characteristics, algorithms of psychophysiological functioning of any one of the representatives of the biological type *Homo sapiens*; of all representatives of this biological type, who currently live on this planet, as well as of all those, who lived in the past or will live in the future. It describes them regardless of gender, race, nationality, place of birth, place of residence, social status and other factors, which are minor compared to the innate structure of psyche. That is, this Catalog describes the entire population.

In the decrypted version, the Catalog of Human Population provides quite a detailed description of any person in terms of his inner, psychical structure in the same way as encyclopedias describe representatives of certain species or subspecies of mammals, birds, fish, plants, etc.: what qualities and characteristics of personality a particular person has, how he thinks, what he is interested in, how he functions in terms of his physiology, what he eats, what kind of emotions he expresses and how, his sex life, his attitude towards family, children, parents, and other people in general, what he strives for, what he seeks in life, his functioning in the environments, his talents; as well as that what he usually hides from others—true goals and objectives, true attitude towards something or someone, preferences, desires, passions and predilections.

You might be wondering: how is this possible? This became possible due to that by nature *Homo sapiens* is a bio-robot, originally programmed living natural object; exactly the same as all other natural objects on this planet, each of which, as it is known, is born strictly programmed for certain conditions of existence, functioning, qualities and properties, way of life, etc. *Homo sapiens* is the same bio-robot as the planet Earth itself and the

Solar system as a whole. After all, any planetary object has its own orbit (from the Latin *orbita* - track, road, path) and a celestial body is unable to get off it; at least not without catastrophes.

It has long been noticed that all elements of this natural system are arranged according to a single principle. Every natural object of the Solar system has its own orbit, either in the literal sense of the word, like the Sun, Earth, Jupiter or the Moon, or in a figurative sense, like animals, plants, and, as it turned out, humans as well. Therefore, it is possible to state that the Catalog of Human Population describes individual "orbits" of *Homo sapiens* in great detail. Each *Homo sapiens* born at any time and at any place goes into one of these "orbits" at the exact moment of cutting of the umbilical cord, which connects him with his mother, and the description of this "orbit" can be found in the Catalog of Mountains and Seas; hence, the possibility to find out absolutely anything about a particular person.

However, since it is not common to use the word "orbit" in reference to natural objects that are not on planetary scale, in the case of animals or humans it is more adequate to use the term "biological program" or simply a "program." Human psyche or, to put is simply, his soul turned out to be nothing other than a program.

The ancient Chinese monument Shan Hai Jing (Catalog of Mountains and Seas) contains two hundred and ninety three descriptions of natural individual (subtype) programs—according to the number of subtypes, which, as it turned out, exist within the "human" biological type. From birth each *Homo sapiens* belongs to one of these two hundred and ninety three subtypes. At the moment of birth, one of the two hundred and ninety three programs turns on in the newborn and a person lives according to this program his entire life.

It does not matter whether an individual knows that he is a bio-robot or not, whether he agrees with this or not, whether he knows his natural program or only vaguely suspects that his life is predetermined from birth. In any of these cases, each *Homo sapiens* lives and functions strictly within his biological program. And, he is not able to go beyond this program; again, not without catastrophes, only in this case not on the cosmic scale, but catastrophes for his personal psyche and physiology.

Any natural object on planet Earth exists and functions solely on the basis of an innate program and is unable to go beyond the limits of qualities, properties, algorithms, functioning set by this program. And, if you think about this, it is not surprising that a human, as part of nature, is not an exception to this rule. After all, what would happen if planets would not have orbits and moved in space randomly, cows would fly, if they wished to, and one day trees would decide to eat animals?

It is very likely that this principle is behind vague philosophical "macrocosm and microcosm are one"—*Homo sapiens* functions according to the same principle as any other object of the Solar system; and maybe of the whole Universe, who knows.

In summary, the answer to the question of how it is possible to get information about any person from a book is as follows: from the very moment of birth, *Homo sapiens* begins to live and function as it is recorded in his natural program (that is, in his soul, psyche), while Shan Hai Jing is the collection of very detailed descriptions of how this happens in the case of each specific individual, as a representative of one of the two hundred and ninety three human subtypes. Shan Hai Jing (Catalog of Mountains and Seas) is the collection of descriptions of human programs.

Homo sapiens, like a computer, cannot function without software. Therefore, using modern computer language, Shan Hai Jing is the collection of detailed descriptions of different types of natural "software" of *Homo sapiens*. Hence, by reading Shan Hai Jing and learning which natural program is implanted in a particular individual from birth, it is possible to obtain any kind of information about this person, find out his true nature without embellishments, as well as that what people were trained to conceal, hide behind numerous social masks and roles.

Having information about the natural individual program and natural individual manipulation modes of a particular person makes it possible to use this information to get your own benefits, goals and objectives in relation to that person, to model his behavior and to influence his psychophysiological state. In essence, this is the answer to the question: "What is the Catalog of Human Population needed for?" Even though this is a very primitive explanation, nevertheless we hope that everyone will understand it. You will be able to find out more about this in the next part of this book.

*****

This book consists of three parts, which, on the one hand, have a single theme, but on the other hand, are devoted to different aspects. In the first part (titled About The Catalog Of Human Population. Introduction To The Topic Of Our Scientific Research) key questions related to the natural "software" of *Homo sapiens* and the Catalog of Human Population, which describes this "software," are presented in a compressed, but at the same time quite detailed form.

In the second part of this book (titled 300 Selected Research Topics Of The Special Scientific Info-Analytical Laboratory—Catalog Of Human Souls (1974-2014) readers can get acquainted with areas and diapason of research topics, which were developed by our laboratory between 1975 and 2014. Of course, these are far from all the topics that we developed and are developing; we listed only those, which, from our point of view, are closest, understandable and relevant for our readers.

This list shows the range of questions, which we not only researched, but also studied very well. As it is easy to guess, it is impossible to provide the details of each one of three hundred topics in the format of one book since it is a huge amount of information. However, if after reading this list someone will want to get detailed information on any of these topics, one can do so by contacting the Special Scientific Info-Analytical Laboratory—Catalog Of Human Souls or the Human Population Academy. Information about how to contact us can be found in the Preface. Also, some of the topics are detailed in our books, the list of which can be found at the end of this book, after the Bibliography.

It should be noted that not all of the topics listed in the second part of this book are related to research of Shan Hai Jing and other ancient monuments. For example, questions concerning artificial "software" of *Homo sapiens*, presented in topics related to women's psychology, psychology of parent-children relationships are not related to the Catalog of Human Population. The ancient Chinese monument Shan Hai Jing does not contain information about artificial "software" because it has nothing to do with natural structure of a human. Artificial "software" was invented by people and, using the computer terms, it is simply a virus and not the basis of human existence, which is the natural "software."

The topic of artificial "software" is described in detail in the third part of this book titled Ahnenerbe—Your Killer Is Under Your Skin. This part describes what it is and what upbringing of children by parents really is in civilization without the Catalog. "Ahnenerbe" (from the German - *Ahnenerbe*) is translated as "ancestral heritage" and in the third part of this book you will learn about what heritage was passed on to you by your parents and mentors.

Ahnenerbe—Your Killer Is Under Your Skin is also published here as an example of one of the topics, which are listed in the second part of this book. This was done to ensure that our readers will not doubt the fact that we really did develop topics that are listed. We have never heard that any of our colleagues in the scientific community had a similar number of research topics and in such a wide range of areas; from this it follows that our readers, in principle, might have such doubts.

In addition, from our point of view, knowing about artificial "software" will be interesting and useful to every reader, as it is about him personally. Many readers are not familiar with the topic of natural "software" that is described in Shan Hai Jing, while every person is familiar with artificial "software."

Therefore, it will not be difficult for a reader to see those processes and mechanisms, which are described in the third part of this book, in his own life and in lives of people around him. This will help readers confirm that Non-Traditional Psychoanalysis, among other things, has sufficient information about both types of "software" of *Homo sapiens*: natural and artificial. We think that it will be beneficial to know this.

*****

In 2014 it was forty years since the start of research by Andrey Davydov—the author of the scientific discovery of the Catalog of Human Population. For over twenty years information from this Catalog was actively being used in practical lives of people. However, those who find out about this, usually ask the following question: "If the Catalog really exists and if it really is such a valuable resource as researchers state, then why it is not being talked about in the media, scientific journals and magazines? And, why is there such conflicting information about this Catalog and its researchers on the Internet?"

These questions are quite natural and we think they need to be answered. However, since the moment of its discovery, the Catalog of Human Population already passed through a certain path and it is necessary to at least briefly tell the story of introduction of the Catalog to the practical lives of people of this civilization. In parallel, this story will be the answer to all the questions mentioned above.

First of all, it is necessary to state that throughout nearly 20-year history of practical use of the Catalog of Human Population everyone, who found out about this discovery, conditionally speaking, divided into two camps; although these two camps were not of equal size.

The first group consisted of different categories of the population, who, after learning about the Catalog, by all means tried to fence themselves off from the fact that a human is a bio-robot and that as a consequence of this now anyone can get the same kind of detailed information about them, as can be obtained about, for example, any of the birds from a professional reference book on ornithology; including that what they carefully hide from others. And also, from that now anyone can control their behavior and influence their psychophysiological state anywhere and at any time, and, most importantly, now a manipulator will succeed at this.

Probably due to horror, these people decided to act according to the principle of an ostrich that buries its head in the sand when there is danger and so they "dived" deep into their fantasies, where their soul remains an unknowable magnitude, "a secret behind seven seals." However, this time such "diving" did not help them. Since the majority of those, who found out about the discovery of the Catalog of Human Population turned out to be people, who are absolutely not fine with the prospect of begging for the rest of their lives: "Lord, enlighten me", "God, give me ..." They were more comfortable with the opportunity to know about themselves and about any person of interest from their environment, and to use this knowledge as they want and have that what they need, including from other people without turning to supernatural powers.

Therefore, while one group remained in fantasies about their unknowability and demanded proof of existence of the Catalog of Human Population from so-called "authoritative sources" in order to somehow justify and comfort themselves, the other group began to actively acquire information from this Catalog. And, not only for the purpose of self-discovery; they were interested in information about other people and application of this information for their own gain. After this "the bold explorers of the unknown" began to actively use all information obtained from the Catalog of Human Population on practice, including to manipulate other people, among which often were "unknowable" fans of mysteries and admirers of "authoritative sources."

And, it should be noted that they were very successful in this, as we personally observed some of the cases from this category because one of our rules is to keep in touch with those, who purchased our research product and take interest in how people use information from the Catalog of Human Population. Observing the processes mentioned above, we found them quite interesting and sometimes amusing.

Any person can be among subjects of manipulation using information from the Catalog of Human Soul (non-scientific title of the Catalog of Human Population). And, as many years of practice of use of the Catalog of Human Population on the territory of Russia (former USSR) have shown, manipulation with information from the Catalog of Human Population occurred regardless of the high social status of subjects of manipulation. This is not surprising since this type of manipulation is absolutely safe for the manipulator because the subject of manipulation is not able to detect this kind of influence and therefore cannot counteract or take counter measures.

It is very easy to manipulate another person when the manipulator has information about him or her from the Catalog of Human Population. And, it does not matter whether the subject of manipulation is "a mere mortal" or, for example, a president of a country—the process of manipulation and the level of difficulty of this process do not depend on the social status of the subject of manipulation.

*****

We think that it is not difficult to guess that besides the developers of the Catalog of Human Population, their colleagues, relatives, friends and acquaintances, as well as those, who managed to find out about its existence from publications in newspapers and magazines, appearances on television, from the Internet—the Catalog was also used by those, who in this civilization are called "powerful people." This is not surprising because they are always the first to learn about everything. The following saying exists not for nothing: "Who masters information—masters the world."

At some point, one of those "great and powerful people" realized that just as easily and freely as he is now manipulating others using the Catalog of Human Population—one day he will be manipulated. And, instead of studying his own natural program and manipulation modes, and through this process becoming immune to any kind of influence from the outside, as others did, this person decided that for returning the Catalog of Human Population to this civilization researchers should not be awarded, for example, the Nobel Prize, but instead be killed. Everything necessary in order to do this was available to this bastard because of his belongingness to the highest ranks of the Russian security services.

We are referring to a group of employees of the Federal Security Service of Russian Federation (former KGB) at the time (2000) led by the colonel of the Foreign Intelligence Service—Andrey Dmitrievich Polonchuk, who has repeatedly turned to Andrey Davydov to obtain information from the Catalog of Human Population about different high-ranking individuals from his organization and from others; including about very high-ranking individuals). However, here we will not tell this story in detail, as well as the story about attempts to kill developers of the Catalog of Human Population. The fifth book in this series is dedicated to this and it is titled Shan Hai Jing—A Book Covered With Blood. The Story Of Developers of the Catalog of Human Population.

Here we will say only that the stupidity of such "polonchuks" trained in barracks of security services did not allow them to come to a simple thought that without having the decryption technology of Shan Hai Jing, without having experience working with this source and other knowledge that developers of the Catalog of Human Population Andrey Davydov and Olga Skorbatyuk and their colleagues have, the possibility to obtain reliable information from there will be zero. They found out about this only fifteen years later.

However, the point is as follows. In addition to attacks that were made on health and lives of developers of the Catalog of Human Population, abovementioned group of FSB officers contributed a lot to casting slurs and slandering developers Catalog of Human Population everywhere possible, as well as to blocking all ways of distributing information about the discovery of this Catalog. Their goal and super-task was to ensure that no one except them would know about the Catalog of Human Population and that no one except them would use this Catalog.

Blocking absolutely all channels of transmission of information (in foreign countries as well) about the scientific discovery made by Andrey Davydov, employees of Russian security services flooded the Internet with plain disinformation about the Catalog and its developers. Therefore, if in any of the Internet search engines one types in "Catalog of Human Population", "Andrey Davydov" or "Olga Skorbatyuk" (especially in the Russian language), then such a huge bucket of stinking slops pours out that one cannot help but wonder: is there such a mutation in nature when the function of the mouth of "*Homo sapiens*" gets replaced by the function of the anus?

However, at some point by some miracle the developers of the Catalog of Human Population managed to appear on Russian television a few times and tell about this discovery. (This was back in 2005 and 2006, and recordings are available at http://www.humanpopulationacademy.org/olga-skorbatyuk-russian-television/.) However, after this access to television audience was cut off for them forever; together with the possibility to contact the scientific community (and not only Russian) and publish information about the Catalog of Human Population anywhere and, first of all, in the media. After all, media is the loudspeaker of those, who hold power and all outlets are tightly controlled, thus media did not have and does not have possibilities to tell about the actual events that occur.

Probably for the same reason as with appearances on television (that is purely "to divert attention," so that it would not occur to anyone that information about the discovery of the Catalog is being blocked by Russian security services), in 2005 Andrey Davydov and Olga Skorbatyuk were allowed to publish several books devoted to the Catalog of Human Population. However, immediately after publication, the first half of printed copies mysteriously disappeared from the warehouse, and with the second half the authors constantly experienced some unexplainable difficulties and problems when they tried to submit them to bookstores in Moscow. After that publications about the Catalog of Human Population on the territory of Russia stopped, even though they took place in the past (and are available at http://www.humanpopulationacademy.org/publications/) before FSB found out about the discovery made by Andrey Davydov.

As for publications about the discovery of the Catalog of Human Population in other countries of the world—this possibility was completely closed due to total blocking of all means of communication for developers of the Catalog: their phones, postal shipments, the Internet.

Through complete blocking of all contacts with the media and access to a wide audience together with the scientific community for developers of the Catalog of Human Population, employees of the Foreign Intelligence Service of the Federal Security Service of Russian Federation led by A.D. Polonchuk completely cut off the opportunity to learn about the discovery of this Catalog for the majority of people living in different countries. Although it is an obvious fact that humanity needs this source of knowledge, and among those people, who found out about the discovery of the Catalog of Human Population the majority were mentally sane and preferred working, practical technologies and recipes over cheap horoscopes, primitive psychological tests and vague philosophical-religious-esoteric ravings.

*****

As a result of events briefly described above, currently not everyone in the world knows that the Catalog of Human Population exists, even though all this time developers of this Catalog and their colleagues tried to tell humanity about the discovery of this unique source. However, another thing is worse. Persecution of developers of the Catalog continues to this day. Specifically for this reason, we do not have offices, where one could come and meet with us, for security reasons we are forced to refuse to publicly provide our phone numbers and addresses, and so on. Also, we are still experiencing difficulties with publishing any kind of information about the Catalog anywhere and everywhere.

We will uncover most recent events as examples. An article about Andrey Davydov—the author of the scientific discovery of the Catalog of Human Population—was created on Wikipedia.org in September 2014. In May 2015 it was suddenly attacked by strange subjects. After they spit out a portion of, to put it mildly, their "waste of brain activity" into a discussion—this article was deleted even though the article did not violate Wikipedia.org's policy. Confirmation of all this can be found at http://en.wikipedia.org/wiki/Andrey_Davydov.

However, our laboratory's staff was not surprised by this incident. And, here is why: the scenario was painfully familiar, as cases similar to this one already happened more than once or twice and some of them are even recorded in Andrey Davydov's political case, which can be found in the abovementioned fifth book of this series (Shan Hai Jing—A Book Covered With Blood. The Story Of Developers of the Catalog of Human Population).

This scenario is banal and primitive to a disgrace. Namely, some supposedly "respectable users" of a resource, ignoring any rules, logic and common sense, as the saying goes, "out of the blue" try to pick on, blame, humiliate, denigrate, slander provided information, and then everything about the

Catalog of Human Population gets removed from the resource. All this gets presented as "an expression of the people's will" in the face of "independent experts" and other "uninterested parties." At the same time, these people behave rudely, absolutely disrespectfully, like louts; or like a farmer, who, without removing dirt from under his fingernails, drank too much, and decided to evaluate and criticize the work of a university professor; or like officers of security services.

The latter seems more believable. Although it is unlikely that we or anyone else will be able to get proof of this—security services do not work like that. Another matter is that officers of security services have the same problem with education and intellect as "ordinary citizens"—instead of work of intellect they have standards and patterns. Any, even the best agent "007" can be identified through these specific patterns and standards by those, who are familiar with these standards well; like we are, for example.

During ten years of forced communication with this kind of dregs we already learned all their habits, "frills and jumps" by heart. And, we can state with certainty that the style of work of these "ordinary users" from Wikipedia.org fully coincides with the style of work of boys and girls from intelligence and security structures of the Russian Federation. They are like a blueprint. And, it does not matter in what language it is being done. Employees of security services are all the same, like cartridges. Their methods of work are standard, they do not think, but act and they act only in the way that they were taught at an academy. It is not very difficult to identify any of them through these methods.

The Internet is flooded with similar subjects. On any high-traffic website "live" a multitude of such "ordinary users" with flawless profiles and excellent reputation, talkative and very demanding as far as compliance with all rules and policies of this website (although they themselves do not observe them). Their only goal is to perform duties of officers of security services, who do what they are ordered to do by their superiors.

Attempts to restore the article about our laboratory's research supervisor Andrey Davydov on Wikipedia.org failed, but this is not the most important thing. The symptomatic fact is that this happened right after an article about the Catalog of Human Population was submitted for publication to one of very large and well-known publishing houses.

Around the same time another characteristic event occurred on the Internet. The link to our official page on Facebook.com (http://www.facebook.com/HumanPopulationAcademy) stopped working. Now those, who want to go to this link to learn more about the Catalog of Human Population or to contact us in order to get information from this Catalog, see the following message: "Sorry, this page isn't available." Even though the page itself was not deleted and still functions, if you sign in to Facebook.com with your username and password.

During forty years since the discovery of the Catalog of Human Population there were so many similar occurrences that a separate book is needed in order to tell about each of them. However, the essence is this: we think that in the situation that developed it hardly makes any sense to puzzle over questions like: "Why the media is not talking about the Catalog of Human Population?" Someone does not want you personally to know about the discovery of the Catalog of Human Population and does not want you to use this Catalog—and that is the "big secret."

It is a pity that a so-called "simple average person" does not understand this very well due to an extreme lack of knowledge about how this society is really arranged and functions. If this was not so, people would not try to accuse us of concealing information about our research and would not demand proof of existence of the Catalog of Human Population from "reliable and authoritative sources," as they would know that "murky water" gets "poured" specifically from these sources.

However, for the majority of population of this planet television is the authoritative source. And, those who have a fancy for watching television programs have quite a low educational and cultural level, interests that do not fall beyond the scope of everyday life, etc. This category of the population does not read books, but reads only price tags in stores and street names on the road, has difficulties adding two and two, and so on. However, most importantly, they are not accustomed to using their intellect and prefer that someone else thinks, researches, tests, and draws conclusions for them.

The majority of average people are unaware of that their transformation into a resource occurred and occurs, first of all, at the informational level; and of that it is enough to convince the masses that black is white and white is black, and then it is very easy not only to "feed" them any kind of informational "stew," but also to do anything at all with them. They are also unaware of that "reliable and authoritative sources," to which people are sent like sheep to a "water-hole," were created specifically for this.

However, these are not our problems, but theirs.

*****

All these years those, who hold power, against our (the discoverer and main developer of the Catalog of Human Population Andrey Davydov and his colleagues) wishes, prevented knowledge about the Catalog of Human Population from spreading to the masses. They did not want and do not want ordinary people to use this Catalog just like they use any other catalogs and encyclopedias—catalogs of cars, real estate, clothing, tools, medicines, encyclopedias of plants and animals, car navigators, anatomical atlases, periodic table, and so on.

Although in essence, there is no difference between the periodic table that describes chemical elements or anatomical atlas that describes different organs and systems of the human body and the Catalog of Human Population that describes *Homo sapiens*. They are all catalogs, reference books, encyclopedias, as the body of knowledge, information on a particular topic.

And, what if today humanity would not have access to the periodic table or anatomy atlases? The answer is clear: humanity would continue to live exactly as it lived during the Stone Age. The same is observed with factual absence of the Catalog of Human Population in this civilization. Not having any idea about the structure and functioning of human species at the level called "psyche," in this segment of knowledge humanity is not at all different from their remote ancestors, who ran after a mammoth with a stone ax.

The consequences of this are quite tragic: there is no normal communication between people, there are no individual recipes for each person (no matter if related to nutrition, philosophy, sexology, etc.), and therefore people are simply unable to engage in anything except attempts to solve problems that arise out of this.

This "civilization" offers a single "one for all" principle, as well as a small number of quite primitive variants, from which each person must find more or less suitable ones for himself. And, he must also find himself in any of the existing social systems on his own. In other words, a person is offered to do that what cannot be done. The results of such practice are sad—while being able to manage most complex technology, *Homo sapiens* is unable to control himself, his psychophysiological processes. In civilization without the Catalog of Human Population, his entire life a human struggles to solve problems, which are not worth a straw, and sometimes without getting any results.

For example, consider an elementary, but a very common problem such as being overweight. This problem is turned into a complex one and does not always get resolved successfully since an individual, not having any idea about his natural program, does not even know what he should eat and, as a result, he eats that what is advertised in supermarkets and on television or that what his mother trained him to eat. And, this is always the cause of obesity and illnesses.

From encyclopedias about animals a person knows that each animal requires its own food, which is different from food that other animals require. And, he knows what will happen if one feeds, for example, hay to a tiger and meat to a horse. However, a human does not know what he himself should eat. This is a paradox. Any deviation from nutrition that is implanted in the natural program of an animal always results in problems ranging from obesity and health problems to death. In this sense a human is not any different from an animal.

Relationships between people in this civilization are also at the level of Neanderthals: figuratively speaking, hit someone's head with a stone ax and no more problems. Misunderstanding, irritation, aggression, hatred, loneliness are problems in interpersonal relationships, which absolutely every person experiences to a lesser or greater extent. Even though, in majority of cases, the only reason of hate towards "thy neighbor" is simply that the other person is different, lives in a different way and is simply unable to live in any other way. However, in order to understand this it is necessary to have the Catalog of Human Population, but people do not have this Catalog, or rather they simply do not know about it.

In addition, the society, throughout the entire history of its existence without the Catalog of Human Population, not being able to properly and effectively control and organize life of *Homo sapiens*, thinks up of newer and newer methods of murder since physical destruction is the only tool that this civilization has in order to influence anyone. And, these methods are successfully implemented. In this civilization human life is not worth a broken penny, even though the opposite is being declared on every corner. Partly for this reason overpopulation of this planet is also seen as a problem in a civilization without the Catalog.

As a result, instead of living happily and developing, *Homo sapiens* is forced to solve a multitude of problems, which would not exist if he had the Catalog of Human Population. In parallel, since birth any member of this society falls into public "millstones," which, after he gets used, mercilessly mill him. And, in this sense scientific and technological progress cannot save a person in any way.

Anthropologists, psychologists, sociologists, medics still continue to rack their brains over not only shocking manifestations of human potential, but also over elementary question of inner differences of one person from another. Although anyone, who is familiar with the Catalog of Human Population knows very well that all these differences are due only to that by nature the "human" biological type is divided into two hundred and ninety three subtype structures, whose inner structures are very different, even though they look similar on the outside (two arms, two legs, lack of scales, tail, horns, etc.); and therefore, they function differently.

Specifically these differences, inner, invisible to the eye are described in the Catalog of Human Population. Meanwhile, psychologists continue to argue about what is primary—psyche or physiology. They do not even know that psyche is primary, not to mention that they have no idea about what human psyche is, what its structure is and on what basis it functions.

*****

The state of affairs in psychology, anthropology, sociology and other sciences about a human partly answers the second most favorite question of the public to developers of the Catalog of Human Population and our colleagues from the laboratory. That is the question about the reason why they do not have close ties with the scientific community. And, we would like to use this opportunity to answer this question.

Children's game "with science" remained in the distant past for developers of the Catalog of Human Population. In the past, which included work at leading scientific and academic institutions of Russia and cooperation with well-known scientists; for example, with A. E. Lukyanov (a sinologist, scholar of Oriental studies, professor, and Head of the Center of Comparative Studies of Civilizations of Northeast Asia at the Institute of the Far East of Russian Academy of Science), V. V. Malyavin (Russian scholar of Oriental studies, Doctor of Historical Sciences, professor) and a Russian scientist, one of the sons of a party leader G. M. Malenkov well-known during Soviet times. Of course, this is not a complete list.

The developers of the Catalog also worked at scientific institutions, for example, under the supervision of the abovementioned A. E. Lukyanov, and presented at scientific conferences and round table discussions (some of these we published in the fourth book of this series under the title Non-Traditional Psychoanalysis. Selected Scientific Articles And Presentations at Conferences. However, at some point it became more comfortable to sit at our own tables instead of these "round tables."

Also, experimental verification of how the technology of decryption of Shan Hai Jing created by Andrey Davydov works took place in Russian scientific institutions. Experiments included audiovisual division between research subjects and the analytical group, which used the Catalog of Human Population, and those, who communicated with research subjects independently and worked on the basis of existing psychological methods. Results of experiments made it possible to draw conclusions. First of all, information obtained by testing covered psychophysiology of the subject of research at the same time instant and not in any one of phenological cycles. Secondly, there was no classification grid in distribution of information received by psychologists. Thirdly, there was significantly more information obtained using the Catalog of Human Population compared to information obtained by the method of testing and there was data that did not relate to psychology. Fourthly, as it turned out, research subjects themselves belonged to different groups: 1—those who never wondered about patterns of their functional diapason; 2—those who wondered about this question when necessary; 3—those who considered this question completely unsolvable in principle. Proceeding from this, sources of identification of received information using the Catalog of Human Population could be permanently realizable physiological desires of research subjects, which they themselves highlighted as vital and top-priority, as well as data from the group that communicated with them with the goal of testing. Comparative analysis of information obtained from experiments showed that vital psychophysiological needs of test subjects completely matched data obtained using Shan Hai Jing, were supplemented by this data about seventy percent and were perceived as a guide to desired actions. To some extent there was a match with materials of testing, although there was significantly less volume of information obtained in the latter and they were formulated somewhat amorphously, which in general is characteristic of psychological products of similar kind.

However, after experiments took place, the scientific environment was happily forgotten. Decryption of Shan Hai Jing and other ancient sources turned out to be more interesting than earning scientific "medals," and then shaking them along with the rest of the "poodles" at this peculiar "dog show." Developers of the Catalog of Human Population and their colleagues from the laboratory decided that it does not make sense to play science when they can actually engage in science, and do so absolutely autonomously. And this, as it turned out, saves a lot of time and energy, and provides absence of any losses.

However, this does not at all mean that developers of the Catalog of Human Population Andrey Davydov and Olga Skorbatyuk and our other colleagues from the laboratory currently do not have any contacts with the scientific community. It is just that now they have those contacts and in that form, in which they are necessary and convenient for them. Moreover, we are open to new contacts. No one ever objected or objects to this. All of us, not only the developers of the Catalog, objected, object and will object to one thing only—participation in processes that are meaningless in terms of what we are engaged in.

*****

In conclusion, we will allow ourselves to make a joke since this book is devoted to the anniversary.

Though this joke might seem unfortunate to someone, but we congratulate the majority of the population of this planet with that for already forty years they do not know about the Catalog of Human Population. And, this means that they still do not know who they are real, what they were born for and what they live for; do not know that in the time period between their birth and death they do everything and anything except be themselves and live their own lives; that they suffer when they could be happy; that they get sick when they could be healthy; that they need money when they could be rich; and that they die when they could live. And, all this occurs in spite of that a source exists right next to them, which has all the necessary recipes for a healthy, wealthy and happy life for each one of them.

Seriously though this humorous congratulation sounds sad and for everything mentioned above humanity should thank not us, but this "civilization" that raised mutants; mutants, who cut off the possibility for the majority of people living on this planet to know about the existence of the Catalog of Human Population, to use this Catalog and to live happily and well, and to become those, who they were intended to be—humans.

Another question is that humanity successfully misled by Charles Darwin about that Homo sapiens evolved from apes identified with animals so long ago, became so accustomed to the animal way of life that possibly some are quite satisfied with this way of life. After all, the most important thing is not to torment yourself with thoughts and questions, right? Let someone else do it for you.

However, from our point of view, the root of people's problems in this civilization is precisely in that they were instilled to believe that someone else should think for them and that someone else must arrange a prosperous, interesting and meaningful life for them. Probably, from here come tales about the Savior, who will come and solve all your problems at once. People wait for the Savior all their lives, but he never comes. Paradoxically, the fact is that this does not bother anyone, people continue to "believe, hope and wait."

However, nature arranged it in another way—no one besides you yourself will create the most comfortable and favorable environment for you, human. Either you do it for yourself or someone else will do it for you, but this "someone" will do this taking into account his personal benefits, not yours. And, this can easily be seen everywhere. The environment gets created for *Homo sapiens*, thereby turning them into a resource; a resource of any kind: labor, financial, intellectual, creative, self-reproducing, and so on.

"*Homo sapiens*" were convinced that each one of them is wonderful, valuable and meaningful, although they were not told that they are wonderful specifically as a resource. They were convinced that society will arrange the environment for them for their well-being purely for humanitarian reasons. And, people believe this, even though humanism simply has nowhere to come from in a civilization where not a single human exists.

Society provides answers to a person's questions about who he is, what he is like and why, and creates the environment for him, but an average consumer pays a huge price for his. Firstly, an opportunity to become a human closes for him. Secondly, he no longer belongs to himself. He entirely (along with his intellect, arms and legs, kidneys, liver, heart, etc.)

belongs to society. And, from this moment on all his psychophysiological wealth will be used as it is advantageous to organizers of this society. Thirdly, an individual will never become a human. Until the end of his life he will exist and function at the level of an animal, or to put it less gently—at the level of domesticated cattle. Favorable conditions get created for cattle as well, but then steaks or wool socks get made from it.

However, in order to create an environment for yourself on your own and to be your own master, it is necessary to do this on the basis of knowledge, not your fantasies; knowledge about the world, about nature as the basis of human existence, about society, etc., but most importantly, on the basis of knowledge about your own natural structure. Since, as mentioned above, if you do not have knowledge of who you are, your structure and address questions like "Who am I?", "What am I like?", "How should I live?" to society—after all, the society just might answer these questions. However, these answers will be beneficial not to you personally, but to someone else. In this case, the master of life of a person becomes the one, who provided all this information to him.

However, the main deception probably is in that all of you were convinced that your fantasies (in which you stay day and night with such pleasure) are your individuality, your essence, your "holy of holies" and you desperately defend them—this is your soul. However, as it turned out, the soul is completely different. Soul (psyche) is natural "software," given to a human from birth.

By the way, human intellect, to which human fantasies belong, meaning the sphere of imagination is not psyche, as it is often considered by psychologists. Intellect is only a part of psyche; it is one of six factors, of which psyche consists. And, normally intellect is exploited not only as a sphere of imagination, but in order to communicate with one's natural "software." However, that is a different topic.

All of you were also convinced that by being in your fantasies (or, using your language, "in the space of your soul") you are in a private place, which no one can enter without your permission and that you are absolute safe there. In vain, as precisely through the sphere of human imagination where images live, it is possible to affect and influence him in any way and with any consequences.

You personally do not know about this since "authoritative sources" did not tell you about this. However, those who do this to you know about this very well. All of you are intimidated by zombiing, but you do not really know how this process occurs technically. Therefore, probably, it is a secret for you that a zombie is not someone out there, but you personally. Those, who are interested, can read about this in the third part of this book titled Ahnenerbe—Your Killer Is Under Your Skin. We just wanted to let you know that you were cruelly deceived, when you were forced to consider your fantasies your soul. Moreover, your fantasies are not even your own.

If a person does not know his natural "software," using the computer terms, artificial programs get "installed" in him. And, through these programs it is possible to make a person live in any way that the "programmer" needs. All that needs to be done in order for a virus in the form of artificial software to start working in you and begins to turn you into a zombie, a nothing or even to kill you is to "fall" into your fantasies. Or a daydream, for example, in front of a television, as you all love doing so much. As you can see, it is very easy to turn any person into a one hundred percent zombie. And, unfortunately, you are one of them.

However, do not rush to cut off your eyelids, as the founder of Buddhism did, according to the legend. Everything is much easier and without self-maiming. In the sphere of your imagination (fantasies) should live images of your natural program, instead of those images, which you see on television screens that you so dearly love, on covers of glossy magazines, web pages, advertisements along the roads or in supermarkets, or which get through the ear canal from anywhere, for example, from your favorite hits. Then, everything will be fine with you. Especially since today it is not a problem to find out images of your natural program. For this it is no longer required to destroy your psychophysiology with drugs or to sit facing a wall for nine years—all you need to do is look in the Catalog of Human Population, in which any kind of information about any person is available.

*****

However, on the basis of everything stated above, there is no need to consider us gurus or even worse revolutionists, who encroach on foundations of this civilization. It does not make sense to encroach on this civilization—it has long been dead. And, it is necessary to admit that death suits it, as this civilization is without a human.

We state that this civilization is dead on the basis of those facts, which are detailed in one of our other books titled Ideology of Religions: Scientific Proof Of Existence Of "God"—The Catalog Of Human Population (by Andrey Davydov and Olga Skorbatyuk). However, there is no need to suspect that we are religious. We are not believers, but you are. All people in this civilization, who do not have knowledge, are believers, regardless of whether they want to be believers or not, know about this or not. Moreover, they are taught to believe, not to know. For thousands of years, on purpose they are trained to believe, and it does not matter in what: in that God exists or in that he does not exist, in science or in secret knowledge, in technological progress or in reincarnation.

In each one of our books about the Catalog of Human Population, our goal is to inform people of this civilization not only about that the Catalog exists, but also about that they can live like human beings, if they will want to. About that they can stop being zombies, animals, if they will want to; because on the inside each one of them has such a great, such an impressive natural potential that it is a pity to see how they live. Therefore, no matter how we perceive this civilization, by returning the Catalog of Human Population to humanity, we think that we are making a huge contribution to that it continues to exist; and, not dead, but alive.

However, the question of whether you personally should use the Catalog of Human Population or not we leave to your own discretion. We absolutely do not want to participate in this choosing. And, there is no need to confuse us with the Second Coming, or with Batman or with heroes of other children's fairy-tales, which you love so much. We have things to do instead of trying to change the destiny of this civilization or to rescue humanity. Moreover, you were deceived in this as well—there are things, which every person must decide for himself, and there are actions, which no one will be able to perform for him; it is just as impossible as it is to eat lunch or ease nature for someone.

Therefore, we are not concerned with how this society lives now or how it will live in the future. We take care only of ourselves, our personal psyche and physiology, our environment, our personal objectives and achievement of our personal goals. Yes, no one can help us achieve our objectives and no one can do anything instead of us, but at the same time no one can prevent us from doing this. And, we must admit that we are immensely happy about this because in this case it is possible to kill us, but it is no longer possible to force us to live as someone wants. And, in fact, this is the difference between a person, who is his own master and a person-resource, who looks into the future with faith and hope, and exists carefree, comfortably and more or less well during a short period of time between pasture and slaughterhouse.

We are ready to inform about that the Catalog of Human Population exists and to provide information from this Catalog about any person to anyone interested. However, if separate individuals or the entire humanity will want to continue existing and functioning at the level of animals—we have nothing against this. We respect any choice.

Civilization can continue to exist in any way it wants—with the Catalog of Human Population or without it. We are simply fulfilling our duty by telling about the Catalog. And, this duty is not before society, but before our creator or creators, who left the Catalog of Human Population in this civilization, probably so that *Homo sapiens* would have the opportunity to live like human beings and to become humans.

As for how humanity decides to use this information—it is not our business. We act according to the principle of a rooster: "I said cock-a-doodle-doo and I couldn't care less if it will dawn or not." Such attitude towards society is convenient for us, as it allows us to take care of our lives and our affairs, and not life and affairs of society. We think that our contribution to civilization is enough because we are working on creation of the Catalog of Human Population, which this population is already using and will use as long as humanity and this civilization exist.

We like the position of the ancient Chinese philosopher Lao Tzu, the author of the famous treatise Tao Te Ching. According to legend, he was stopped by a border guard when he traveled to the West and the guard asked him to share the knowledge of how the world and people are arranged. Lao Tzu accommodated his request and very briefly, in eighty one zhang (a unit of measuring text in China), described ... how to use Shan Hai Jing as the Catalog of Human Population. And, according to our research, Tao Te Ching, I Ching and some of the other ancient sources are nothing other than annotations on how to use Shan Hai Jing.

In his treatise, Lao Tzu clarified what will be with those, who will use the Catalog of Human Population and with those, who will not want to use it. (However, do not rush to read the Tao Te Ching, as you will not know find out what is really being stated in this treatise; in any language of this text, even in the language of the original.) We tried to do something like Lao Tzu managed to do. It is not so important to us whether we did it well or not. The only pity is that we have not yet reached the level of this great man of wisdom, who was able to tell the entire story in eighty one short zhang.

PART 1

ABOUT THE CATALOG OF HUMAN POPULATION: INTRODUCTION TO THE TOPIC OF OUR SCIENTIFIC RESEARCH

Sketch © 2000 Andrey Davydov

UNIT 1

WHAT IS THE CATALOG OF HUMAN POPULATION AND WHAT IS HUMAN PSYCHE?

1.1 Scientific psychology and researcher's, author of scientific discovery Andrey Davydov's finding of the answer to the question "What is human psyche and what is its structure?" in one of the most ancient monuments preserved in our civilization.

People of this civilization have long been convinced that each one of them is unknowable. This is not surprising since it is much easier for "the powerful ones" to control ignorant masses in order to use them. However, *Homo sapiens* was created in such a way that he does not stop being interested in what he is, and how his psychophysiology[1]—the sum of psyche and body—is arranged, despite that this interest could not find satisfaction for many centuries. Religion offered quite an amusing, to put it mildly, vision of these issues, and psychology, until recently, was unable to provide authentic information to the most elementary questions of an individual, such as "Who am I?" and "What am I?" For none of the existing methods, which psychologists have, despite their variety, allowed to look inside the human soul.

Although psychologists tried to find a way out of this situation, from our point of view—these attempts were quite unsuccessful. Contrary to that words 'soul' and 'psyche' are synonymous (from the Ancient Greek ψυχικός "soul, soulful," from ψυχή—"soul"), representatives of scientific psychology separated them without much hesitation. Soul was taken out of the scope of their studies, declared a phantom and left to religion, while under the label "psyche" they began to study anything at all: "phenomena of consciousness", "direct experience of the subject," his behavior, intellect, etc. It got to that intellect was practically announced to be the soul. That is, psychologists are not even aware of that intellect is not psyche, but only a part of psyche. Moreover, it is a secondary part (more on this will be presented below).

All in all, it turns out that the science of psychology was never a science because it never had and does not have any idea about the subject of its study—the soul, psyche. Moreover, it persistently ignores its subject, even though "psychology" translated from the Ancient Greek means "science of the soul" (from the Ancient Greek ψυχή—"soul"; λόγος—"teaching"). Scientific status of psychology not only became the topic of extensive debates, but also is denied by psychologists themselves. As noted by A. V. Yurevich (the deputy director of the Institute of Psychology of Russian Academy of Sciences) in 2005, psychology occupies an intermediate position between science and pseudoscience. And, surprisingly, psychologists are fine with this.

However, despite everything, at the end of the XX century answers to questions "What is human psyche and what is its structure?" and many others—were found. To put it simply: the human soul was opened liked a tin can. Scientists decrypted one of the most ancient monuments preserved in this civilization, and it turned out that this source contains a detailed description of the structure and mechanisms of work of human psyche. (Or human soul, whatever you prefer.)

The name of this ancient monument is 山海經 (*Shānhǎi Jīng*, Shan Hai Jing, translated as the Catalog of Mountains and Seas, Book of Mountains and Seas). Precise dating of this monument is still not available, but some researchers date it XXIII century BC. In the 70s of the XX century, that is—almost forty centuries after the presumptive appearance of the above-mentioned monument in this civilization, sinologist and researcher of ancient books Andrey Davydov posed a question: "What is the Catalog of Mountains and Seas? Is it an unclear treatise or the richest collection of images, which are directly related to a human and his psyche?" This question was not empty because images of strange beings, often of chimerical nature, remained in practically all cultures of the world, but nowhere, except for in the Catalog of Mountains and Seas, these images were systematized into something whole that could be called a catalog. And, Andrey Davydov found the answer to this question. Shan Hai Jing indeed turned out to be a map of psychophysiological structure of *Homo sapiens*—the Catalog of Human Population.

This ancient source was qualified as the Catalog of the human population on the basis of the fact that Andrey Davydov found in it detailed descriptions of 293 archetypal structures of human psyche. This clearly indicated that the biological type "human" is divided into subtypes (subspecies), and Shan Hai Jing describes the psychophysiological arrangement of each one of them, similar to catalogs, encyclopedias about animals, anatomical atlases, etc.

In the course of many years (from 1975 to 1995) of research, Andrey Davydov not only managed to create a unique technology for decryption of Shan Hai Jing, but also decrypted the information found in this source. This information included description of the arrangement and work of psychophysiology of 293 human subtypes, as well as details about the natural toolkit for managing a human in the form of descriptions of natural manipulation modes for each subtype, and a lot of other interesting information related to natural arrangement of *Homo sapiens*.

It became clear that the opinion (which is still widespread) that at the moment of birth a human is "a tabula rasa" ("a blank slate"), on which anyone, including himself, could write anything that he pleases, like on a blank sheet of paper, is a fallacy. Detailed descriptions of 293 structures of human psyche, which were discovered in the ancient monument, clearly demonstrated that every person is born with a natural program, in which everything is implanted: what he will be like, how he will live, down to the smallest details. It turned out that the soul does exist. And, it is not some kind of unknowable value or a phantom. Using the computer terms, the soul (psyche) is the "software" of *Homo sapiens*.

One's own soul or the soul of another is no longer a mystery because from descriptions, which were discovered in Shan Hai Jing, it is possible to obtain any kind of information about any one of over 7,000,000,000 representatives of the biological type *Homo sapiens* living on this planet. However, it is not a mystery only to those, who have access to decrypted information from this ancient source. The rest still have to ask psychologists and society: "Who am I? What am I?" and receive absurdity and someone's imagination in response.

In parallel, psychology got a chance to be called science. After all, owing to many years of research work carried out by Andrey Davydov, answers to questions "What is the soul (psyche) of a human and what is its structure?"

were found. They were discovered in the Catalog of Human Population, which turned out to be ancient writings, heritage of our ancestors or maybe our creators because the authorship of this source is undetermined. Decryption of Shan Hai Jing, as translation from the language of ancient images into the language of modern descriptions of life and functioning of two hundred and ninety three human subtypes, formed the basis for the Non-traditional Psychoanalysis, as a new direction in scientific psychology. This direction factually made it a science.

However, something else is more important. With the return of the Catalog of Human Population to this civilization, every person living on earth (and their descendants) has a chance to become a human. Currently, there is not a single human (according to the meaning of this word, which is recorded in the catalog that describes humanity) present on this planet. Ancient writings clearly state that one is not born a human, but one can become a human with the help of a certain, specific method—learning; learning as studying and mastering one's own natural "software." Moreover, this includes going through multiple levels of development. However, it is not possible to become a human without the "textbook" called the Catalog of Mountains and Seas.

[1] We use the term "psychophysiology" because for us the issue, which is still being discussed by psychologists—"What is primary: psyche or physiology?"—has long been settled. We know that psyche is primary. However, having no relation to religion, we do not separate the body and the soul, and we do not consider the soul as something that exists by itself, apart from the body. And, in fact, this is not our personal opinion—this also is information from ancient texts, which appeared long before Torah, Bible, and so on.

1.2 What is the soul (psyche) of a human and what is the Catalog of Human Population?

Soul (psyche) of a human is an archetypal matrix. Or, if we draw an analogy with a computer, "a hard disk," on which natural "software" of *Homo sapiens* is recorded. Every person is born with an individual program and three manipulation modes; this can be defined as "software," based on which he lives, and on the basis of which all his psychophysiology functions.

Since all research related to the Catalog of Human Population was and is being carried out within a strictly scientific framework, the following is the scientific definition: "The Catalog of Human Population is a description of a human as a type by subtype structures. Subtype structure (synonyms: "psyche", "soul") is a combination of individual archetypes, recorded at the genetic level (principle). Expressions and interaction of subtype structures in manipulation modes and phenological algorithms are described with adjustments for gender, age and cultural differences. Information is recorded on six factors." This definition was formulated and introduced by Andrey Davydov—the author of the discovery and decryption of the Catalog of Human Population.

1.3 *Homo sapiens* is a biological type divided into 293 subtypes.

The basis for cataloging of the "human" biological type was the discovery of existence of subtypes within this type. This discovery was made by the researcher-sinologist Andrey Davydov. It turned out that there are 293 subtypes of *Homo sapiens*. Each of them, using the language of science, is endemic. That is, representatives of one subtype cardinally differ from representatives of any other subtype due to particularities of arrangement and work of psychophysiology, qualities and characteristics, life algorithms, functional range, and abilities.

1.4 How does a representative of one human subtype of *Homo sapiens* differ from a representative of another subtype?

It has long been noted that even though people are physiologically similar in principle (for example, they have two arms, two legs, two lungs, one heart, and do not have scales, a tail, horns, hooves, wings, feathers, etc.), they differ from each other in character, qualities of personality, life algorithms, values, preferences, and abilities. And, sometimes they differ cardinally On the other hand, there are people, who are similar to one another. This question remained a mystery to researchers for a long time, but with the discovery of the Catalog of Human Population it was resolved. It turned out that the reason is that humanity as a biological type is divided into subtypes.

It is exactly the same as it is arranged in the world of animals, birds, fish, plants, etc. Among birds exist hummingbirds, and eagles, and sparrows, and ostriches; among fish there is a shark, and a carp, and a guppy or a dwarf pygmy goby; among snakes there are anacondas, and vipers, and grass snakes; among dogs there are huge Great Danes, and miniature Italian Greyhounds. And, as it is known, anacondas do not live like vipers, sharks like carps, and Great Danes like Italian Greyhounds—they all lead different lifestyles, have different qualities, functionalities, and abilities. As it turned out, it is the same within the biological type *Homo sapiens*.

1.5 How does one individual differ from another, as representatives of the same subtype of *Homo sapiens*?

In principle, they do not differ in anything. Externally, representatives of the same human subtype, unlike animals of the same subtype, might differ, including by the color of their skin, for example. However, they have the same qualities of personality, lifestyle, abilities, and functionalities. Despite that externally representatives of one subtype of biological type "human" usually do not look similar, their internal physiological processes are analogous. For example, if we consider the possibility of normal existence in different temperature regimes, then that what is natural and desirable for a representative of one human subtype might be completely unacceptable to a representative of another, etc.

This occurs because all physiological processes are determined by the structure of psyche, and it is identical for all representatives of one human subtype. Representatives of the same subtype of *Homo sapiens* are born with the same subtype program, and therefore, in terms of psychical structure they are identical. The structure of psyche (subtype program) is a natural imprint, a matrix, and it does not change from individual to individual within a subtype.

Social environment, upbringing, cultural traditions, and education introduce some adjustments, but they are minor since they do not change the essential—subtype structure ("program") of a person. Real differences between people are defined specifically by the natural subtype program, nothing else.

The study of Shan Hai Jing revealed that every person, belonging to some subtype, has stable characteristic properties of this subtype; and, regardless of race, nationality, place of birth and residence, parental guidance, and so on, as these are only minor correctors, which do not change a subtype program. Though, the same pattern is observed with animals. Living in different areas (as some animal subspecies, as well as human subspecies can be found almost all over the globe), the representatives of the same subtype may have different adaptive properties, but these properties do not in any way alter the subtype program: a horse remains a horse, a bear remains a bear, a bull does not turn into a ram, and a snake does not become a crocodile. Only here it will be a pony, somewhere else a percheron, and elsewhere—an Arabian horse. Regardless of the territory where an animal lives, it retains stable properties of its subtype. We see a similar situation with humans: the representatives of the same human subtype might have different skin color, be of different nationalities, but each retains those personal qualities, life algorithms, preferences, talents, functioning that are inherent in his subtype.

Cultural factors, national traditions, even parental upbringing (which, by the way, plays a big role for representatives of some subtypes) are only additions, layers, and do not alter the subtype program of a human. Since, in the case of biological species *Homo sapiens*, children are not direct continuation of their parents, as it is the case with animals. More often than not, children and parents are members of different subtypes.

1.6 Why and in which cases children are not antenna-like continuations of their biological parents?

As it turned out with the discovery of the Catalog of Human Population—since humanity as a biological type is divided into 293 subtypes, having a purely external similarity ("exterior") a child does not inherit natural psychophysiological characteristics of his biological parents, except for when he is a representative of the same subtype as one of his parents (a mother or a father).

In the case of representatives of the biological type "human," children are not antenna-like continuations of their biological parents, as it is the case in the world of animals and plants. In the world of *Homo sapiens* in most cases parents raise a representative or representatives who are not the same as their own subtype. This is the root cause of the well-known "problem of parents and children." This is why children, even though they resemble their parents externally, in most cases do not get all characteristics and qualities of their characters, do not copy their way of life exactly as it happens in the animal world. And, sometimes they cardinally differ from their parents by nature.

As for similarity of characters, habits, attitudes, etc. observed in parents and children—this has nothing to do with their nature, and is only related to artificial "software." We have fully researched this issue and it is described using popular language in the book titled Ahnenerbe—Your Killer Is Under Your Skin by A. Davydov and O. Skorbatyuk (also available in Part 3 of this book).

UNIT 2

WHAT IS HUMAN PSYCHE ("SOUL") AS "SOFTWARE?" WHAT TYPE OF SOFTWARE CAN IT BE?

2.1 *Homo sapiens* is a bio-robot.

Using scientific terminology, *Homo sapiens* is a bio-form that has a program on the basis of which it functions in space and time, just like all other bio-forms on planet Earth. In other words, a human is the same bio-robot as all other natural objects.

A human, first of all, is a natural object, an element of the unified natural system. Any animal or plant lives and functions within strict programs. And, our planet itself and other elements of the Solar system are not able to move in space arbitrarily, deviate from their orbits; meaning that they are objects, which are programmed from the beginning.

A human is subject to the same natural laws. Subtype program of *Homo sapiens* gets activated at the moment of birth of an individual. And, he begins to live strictly according to this program, even if he knows nothing about it. It is the same as it is with, for example, animals and plants, or planets of the Solar system: Earth does not "visit" the orbit of Mars, a fish is not able to arbitrarily begin to live and function like a horse, and an oak cannot stop being an oak and begin to exist like a palm tree.

2.2 *Homo sapiens* is born with built-in "software": individual (subtype) program and three manipulation modes.

In essence, the concept of human "software" (or, using scientific terms—"subtype structure" or "individual archetypal pattern"), "psyche," and "soul" are synonymous. The role of human "software" is huge. It is the basis of work of all human psychophysiology.

Human "software" is the natural subtype program of an individual and three manipulation modes as correctors to this program: suppression mode, balance mode and stimulation mode. A program and three manipulation modes are "soldered together," as elements of a unified system called "human psyche." Every human subtype has its own, unique natural program with three manipulation modes. More information about programs and manipulation modes of *Homo sapiens* will be provided below.

Human "software," using the computer terms, gets "installed" by nature at the time of birth. If we draw an analogy with other technics, an individual program and three manipulation modes of a person are, relatively speaking, "factory settings," meaning that an individual is created by nature specifically like that.

2.3 Natural and artificial "software" of *Homo sapiens*.

A human and a computer have a similarity: both do not work without software. A computer without "software" is a useless pile of metal; a human without "software" is a body, life in which is supported by automatic physiological processes, but he will never function like a human. Today almost everyone is familiar with computer software, but not many are familiar with human "software." Even though, as it turned out, human "software" is the "soul," which people were looking for so hard for centuries, but could not find.

Human "software" can be natural, as well as artificially created by people themselves.

Nature gives "software" to every person from birth as a base, without which any bio-robot, including a human, cannot live. Without even coming close to that natural "software" is provided by nature to every person from birth, civilization came up with a lot of modifications of artificial "software." People had to live somehow since by nature it is arranged so that a person cannot live without a program at all, and the Catalog human population, where every person could find out his natural program was lost.

However, the trouble is that artificial "software" (meaning, "software" invented by people) differs from natural in the same way as plastic apples differ from natural apples. That is not surprising because even at the current level of technological development of this civilization, a human is not able to create anything that would be equal to natural; quality of that what he creates is always worse. Let's go back to "plastic apples": such a "diet" quickly leads to death of a human. In this civilization, in the best case, people live for several decades, instead of centuries, as they could live according to studies done by physiologists.

In essence, life based on artificial "software" that replaces the natural program of an individual, turns him into a living corpse while he is still alive. Life on artificial "software" leads to colossally rapid deterioration of the entire human psychophysiology, premature aging and early death. Only a few percent of the potential that is implanted in a person (and, which would allow him to heal and rejuvenate himself) gets realized and a person has nothing else left, but to die from diseases and old age.

And, during the time period between birth and death, for the most part, a person's life is difficult, full of problems. Is it not this process and result that are described in mythologies of almost all nations of the world: a shift from the Golden Age, in which all people lived for many years, happily and without problems to the Iron Age, tainted and cruel, when "men never rest from labour and sorrow by day, and from perishing by night"[1]?

Artificial "software," like hydrochloric acid, corrodes a person from the inside. As it is known, concentrated hydrochloric acid turns contents of a stomach into a bloody mess. A chemical burn of this kind does not always end instantaneously, sometimes doctors save a person by removing the insides, but a person lives the rest of his life severely disabled and dies before his time. And, at any stage a person is in hellish pain because of this. Is this not the same process that we can observe around us: life in misery, and then death, in 100% out of 100 cases?

No person in his lifetime is able to fall outside the scope of his natural program without damage to himself. And, everything that *Homo sapiens* try to make out of themselves by creating artificial programs, as replacements of their natural programs, in terms of nature are mutations. In nature all mutants automatically "go under the knife." Automatically—means that if functioning of an individual falls outside the scope of his natural program, then nature automatically regards this as a threat to existence of the whole natural system, and the program of self-destruction automatically turns on in that individual.

Probably, this is why lives of *Homo sapiens* in this civilization are so short. After all, according to physiologists—by nature a human body is designed for a healthy, normal and youthful life of a few hundred years. And, this is confirmed by the ancient sources; for example, in descriptions of lives of biblical characters, which most people are familiar with. One of them—Methuselah—lived 969 years according to some sources. Or is the Bible wrong?

Artificial "software" is the root of absolutely any problem of an individual, as well as society as a whole. "Software" created by a human for himself and for other people can be compared to the most dangerous computer virus, which breaks the entire system; to a virus, which there is not a single chance to get rid of, except for doing "a factory reset," cleaning "the disk" called "the soul" of everything extraneous, everything that is not related to the natural "software." And, today this is possible since the Catalog of Human Population, which is the collection of descriptions of human programs, was found and returned to civilization.

Artificial schemes that people come up with as programs for life are unable to completely turn off their natural "software." Simultaneous work of both natural and artificial "software," first of all, makes the life of an individual very uncomfortable, as these two "programs" come into a conflict, which does not end until death. Secondly, this is very similar to trying to sit on two chairs at once. And, the result is the same. Trying to resolve this conflict, but not knowing his natural program, at some point an individual completely switches over to artificial "software." The trouble is that when he "succeeds"—he dies right away; a person is alive only while his natural program works. Therefore, natural "software" of people, which is described in the Catalog of Human Population, always corresponds to real people. And, this is easily observable.

[1] *Hesiod* [VII-VII century BC]. Works and Days. Retrieved March 14, 2015 from http://www.sacred-texts.com/cla/hesiod/works.htm.

UNIT 3

NATURAL INDIVIDUAL (SUBTYPE) PROGRAMS OF *HOMO SAPIENS*

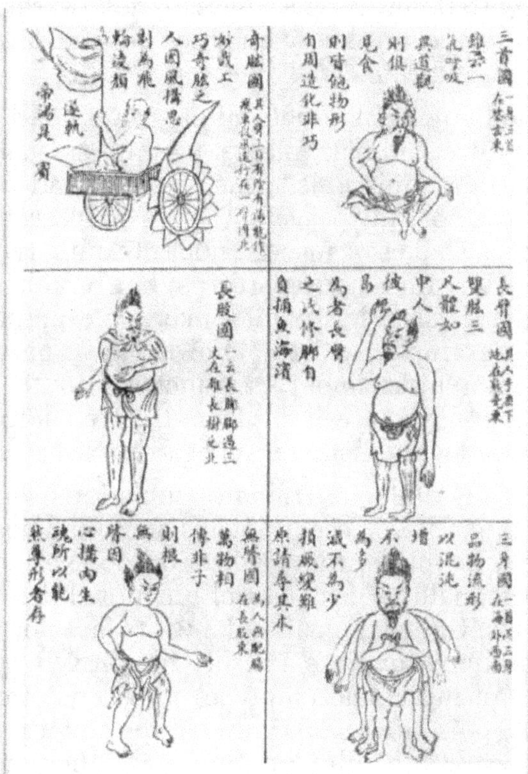

3.1 Why know the natural individual (subtype) program of a human?

Characteristics of species and subspecies of animals and plants, which are different by nature, make it possible to classify and describe them in encyclopedias. This, in turn, allows a person to obtain information about a certain natural object or phenomenon from encyclopedias, reference books and other catalogs. For example, from catalogs of animals—it is possible to learn the natural qualities of each of them, how they look, how they live, what they eat, how they reproduce, etc.

Prior to creation of such catalogs, a human had to learn this "on his own skin." Encyclopedias (catalogs), which describe the world of nature, relieved people from having to learn on their own that some animals can damage or even kill, that plants can be very poisonous, and that water in nature is not only a brook or a calmly flowing river, but also vertiginous currents, swamps that suck in deep, huge ocean waves, floods, etc.

Without an encyclopedia, from which it is possible to easily find out about any person, the same thing happens to those, who without having knowledge about the true nature of their fellow biological species try to come into contact with them. People's lives are full of troubles, dangers and damages because they do not know qualities and functionalities of people with whom they communicate.

After all, as it turned out, human natural programs are a mix of properties of real objects of nature: animals, plants, mountains, valleys, rivers, lakes, swamps, streams, rocks, minerals, soils, metals, and much more. And, among them there are not only peaceful rabbits or sparrows, but also tigers, wild boars, venomous snakes or insects; not only quiet rivers and lakes, but also rapidly flowing streams; not only carps, but piranhas and sharks; not only buttercups and daisies, but plants, like, for example, manchineel or Cerberus. And, for example, without knowledge of natural properties of manchineel, if a person hides under its branches from the rain—drops with mixed-in poisonous milky sap will fall from the tree onto that person, and he will become covered with blisters from head to toe.

Therefore, without the Catalog of Human Population every person has a lot of problems throughout life. These problems never end because he is always among people. And, people in this civilization do everything to hide their true qualities of personality, their natural functionalities behind numerous cover ups and masks. Under the guise of a white lamb or a beautiful flower can hide anyone and anything. As a result, figuratively speaking, a person without knowledge of individual (subtype) programs of other people, first steps on the tail of a poisonous snake and it bites him, then gets sucked into a swamp, then a rock hits him on the head, then he gets ripped by poisonous thorns, then he finds himself in a tornado zone or even paws of a tiger.

For a long time *Homo sapiens* remained one of the few biological objects, which was not arranged systematically by subtype structures in an encyclopedia. Of course, some encyclopedias about a human existed, for example, anatomical atlases. However, one could not get information about what this or that person is like, let alone how to come into contact, interact with him. But now the biological type "human" is classified due to that researcher-sinologist A. Davydov discovered detailed descriptions of two hundred and ninety three human subtypes in the ancient monument titled

Shan Hai Jing, which is why this book was named the Catalog of Human Population.

The Catalog of Human Population is an encyclopedia of *Homo sapiens*, from which it is possible to obtain detailed descriptions of people. From this directory one can easily find out the natural individual (subtype) program of any of over 7 billion people living on planet Earth.

Consideration of each person as a representative of one of the subtypes allows to find out any information about him/her using minimal form data, only by the date of his/her birth. This is very convenient.

The Catalog of Human Population allows to obtain any information about a person on all six factors: what this person is like in the intellectual sphere, how he/she thinks, at what rate and in which algorithms, what his/her interests are, etc.; what, when, where and how he/she eats; what are his/her physical parameters, abilities, etc.; what emotional manifestations are normal for him/her, how powerful they are, how and in which situations he/she expresses emotions, etc.; what are his/her sexual preferences, sexual potential, how it gets expressed, in what forms, with which partners, under what conditions and algorithms, etc.; in what geographical area, type of housing, interior, conditions he/she prefers to live; what his/her natural talents are, how they can be realized (including professional activity, although not necessarily); how he/she communicates with others, on what bases, what he/she seeks, by which methods, and what other people should be fearful of with him/her; and so on.

To know the natural individual (subtype) program of a person means to know all his/her hidden "motivational springs," personality traits, life algorithms, functions, abilities, habits, and predilections; including everything that an individual usually carefully hides from others. This knowledge fully reveals the true nature of a person (which is usually hidden behind numerous masks), giving a complete picture of who he or she is—real, "without embellishments."

In order to make contact with another person and turn it in the needed direction, it is necessary to understand exactly who your opponent is: how he/she will act, what and how he/she will say, what he/she will seek, and so on. To know all this, it is no longer necessary to meet up with this person, to observe him, to try to learn something about him from someone, to provoke some actions, to study his reactions, and so on. There is no longer a need to be in any situation with a person in order to know exactly what to expect from him (what actions, what reactions, what decisions) and what to be cautious of with him. It is enough to simply read the description of a person in the Catalog of Human Population.

It is possible to read any type of information about any person in Catalog of Human Population. It is just as easy as learning, for example, about dogs from a reference book about dogs. And, this is at a minimum because the Catalog of Human Population also contains information about how to make a person absolutely controllable from the outside, about how he should "grow," meaning—gradually develop, self-perfect, etc.

3.2 What is the natural individual (subtype) program of *Homo sapiens*?

Individual (subtype) program of *Homo sapiens* is a program, based on which a person lives. He is unable to go beyond it. A program "starts up" and begins to work at the time of birth; to be more precise, at the time of cutting of the umbilical cord.

Individual program of a person is a program of one of 293 human subtypes. Therefore, we use the term "individual (subtype) program." The program of each subtype differs from programs of all other subtypes.

An individual (subtype) program is recorded in human unconscious by the language of natural images (archetypes). Archetypes are a protolanguage. Unconscious of *Homo sapiens* "speaks" this language and human psychophysiology "understands" only this language. One human program can contain about 20 or over 100 natural images. Individual programs of human subtypes are recorded by different quantities of images, but no subtype has any advantages over other subtypes—all have equal potential and survival rate.

Each individual (subtype) program consists of two parts "active" and "passive."

Each individual (subtype) program is described in the Catalog of Human Population on six factors: intellectual, physical, nutritional, emotional, sexual, and environmental.

Appearance of people with various subtype programs is connected with natural phenological cycles. Therefore, it is necessary to know the day, month, and year of birth of a person in order to find his description in Shan Hai Jing.

3.3 How an individual (subtype) program of a person is connected with his/her date of birth?

It is necessary to know the day, month and year of birth of a person in order to find his/her description in the Catalog of Human Population. However, we have to disappoint astrology and numerology fans, as this is not related to influence of planets and stars, or "magic of numbers." This is related to natural phenological cycles.

For those who are not familiar with the science called "phenology"—in a nutshell, it is a system of knowledge and sets of information about seasonal phenomena of nature, time periods of their occurrences and reasons that define these periods, as well as the science of space-time laws of cyclical changes of natural objects and their complexes, associated with annual movement of the Earth around the Sun.

In nature, absolutely all bio-forms (plants, animals, birds, fish, insects, reptiles, etc.) appear during their time periods. Scientists have recorded this fact a long time ago. At the beginning of the last century, a German zoologist Alfred Brehm has demonstrated that each type of living organism appears on earth during a specific time period.

For example, rabbits are born in late April-early May, deer—in May-June, seals—in March, and so on. "...And the stork in the heaven knoweth the appointed times, and the turtle, and the swallow and the crane observe the time of their coming..." (Jeremiah 8:7) A human, as part of the earth's ecosystem, as part of nature is subject to the same laws as all other natural objects; therefore, if at certain specific time periods creatures, which we call "a cow", "an eagle" or "a hippopotamus" are born, then exactly the same rule applies to a human.

Using scientific terminology, *Homo sapiens* is programmed bio-form, anthropo-zoomorphic being, individual (subtype) archetypal pattern of which depends on phenological cycling of a geographical area. Appearance of subtypes of the biological type *Homo sapiens* is connected with natural phenological cycles, and specifically for this reason, classification of subtypes in the Catalog of Human Population is recorded according to the date of birth.

Andrey Davydov—the author of the discovery of the Catalog of Human Population in the ancient Chinese monument Shan Hai Jing, and the author of creation of the technology of decryption of this ancient monument—put together a schedule of appearance of subtype structures of type human, and this schedule is copyrighted.

3.4 How 293 individual subtype programs are distributed if there are 365 or 366 (leap years) days in a year?

Despite the fact that birth of a particular subtype structure is firmly connected to the date of birth, in the case of research of the ancient monument Shan Hai Jing (which turned out to be the Catalog of Human Population) things are not as simple as they are in astrology. At least because there are 365 days in a year (or 366 in leap year), while there are 293 subtype structures of *Homo sapiens* described in the Book of Mountains and Seas. However, this problem has long been solved by Andrey Davydov—the author of the scientific discovery of the Catalog.

However, many of those who learn about the Catalog are concerned about the question of how 293 individual subtype programs are distributed if there are 365 or 366 (leap years) days in a year. We uncovered that some subtype programs are reproduced a few days in a row. However, this does not mean that people born in this period are absolutely the same because manipulation modes of these individual programs are often completely different. Therefore, to put it in simple terms, these people are not like each other one hundred percent, but they do share the same subtype program—meaning, individual qualities and algorithms.

3.5 When does a human subtype program turn on?

"Hour X," as time of birth of a human, starts up his subtype program. The "soul" begins to work, and a human—to live. And, he will live while his soul, in other words—his subtype program as a mechanism works in his body, and works on all six factors: intellect, physiology, nutrition, emotions, sexual factor and environmental factor. If the work of one of these factors stops, then breakage of other factors will follow: fast, as in the case of, for example, physical and nutritional factors (starvation, disease), or slow, as, for example, in the case of sexual, emotional or intellectual factors.

It does not matter how a person was born, whether in a natural way or, for example, with the use of cesarean section. Individual subtype program starts working at the moment of cutting of the umbilical cord. It does not matter when a person was conceived, how long the pregnancy lasted, and so on. The only important moment is the time of detachment of his body from the body of his mother in the form of cutting of the umbilical cord because that is the moment when the individual (subtype) program begins to work.

There is a supposition (although we have not studied this issue within interdisciplinary research) that subtype programs of *Homo sapiens* is the record that human DNA contains. It is very likely that every human embryo initially contains a record of all 293 subtype structures, along with manipulation modes for each one of them and a record of passage through these stages of personal (subtype) evolution of an individual, which are also programmed and are described in Shan Hai Jing. This is a huge array of information. However, at the moment of birth only one of the subtype structures is activated, and all other records remain inactivated. Once again, unlike everything else presented here, this is only a hypothesis.

3.6. Is it necessary to know the time of birth besides the day, month and year of birth of a person in order to find his/her description in the Catalog of Human Population?

This information is required only when there is data that a person was born close to midnight—half an hour before midnight, or within a half hour after midnight. In this case, special technology is used to determine where the description of this person is located in the Catalog of Mountains and Seas. As for the time of birth within one 24 hour period—this factor does not matter for finding a description of a person in the Catalog of Human Population.

3.7 Why is it necessary to have accurate information about the year of birth of a person?

It is very important to know the year of birth of a person in order to get information about him/her from the Catalog of Human Population. It is important to know the year for one reason only: to understand whether it is a leap year or a common year. Since in most cases if two people were born on the same day, but one of them was born in a common year and the other was born in a leap years, then they are representatives of different human subtypes. Therefore, they have different individual (subtype) programs and manipulations modes. In this case, they have completely different qualities, characteristics and functions, they lead different lifestyles.

Of course, coincidences do happen because the ancient monument Shan Hai Jing contains 293 descriptions of human programs and, as it is known, there are 365 days in a year. However, in order to use this information, it is necessary to know exactly which programs are reproduced multiples times.

Therefore, without having knowledge of the exact year of birth of the person of interest—accurate information from the Catalog of Human Population about this person cannot be obtained. However, this is not a problem because in the modern world it is very easy to get information about when a person was born. Usually, it does not even cross a person's mind to hide this kind of information about himself. Also, it is easy to find it in public records, profiles in social networks, documents, forms, and so on.

3.8 What are "Active" and "Passive" of an individual (subtype) program of a person?

As it turned out, any natural program of *Homo sapiens*, regardless of subtype, consists of two parts. The author of this scientific discovery, researcher-sinologist Andrey Davydov named these parts "Active" and "Passive."

The "Active" part of a human program is responsible for the period of a person's activity. It is a period of real actions, but also a period of active energy and strength expenditure. Therefore, every person requires rest from time to time, and not only in the form of sleep, but also in the form of some type of activity. Not for nothing there exists a saying: "The best rest is change of activity." Human activities from the "Passive" part of his individual program do not take away strength, but rather help a person accumulate it, turn on regeneration processes in him.

Not knowing about the existence of this natural mechanism, which is built into every human, people are often frightened of that two different personalities are always present in others or in themselves all the time. The diagnosis known as "dissociative identity disorder" does not require any treatment, if we do not consider clinical cases in the field of psychiatry when, using the language of Non-traditional Psychoanalysis, a serious malfunction of the individual (subtype) program occurs.

"Active" and "Passive," to put it very raspingly and primitively—are like two different human characters because each part of the human program is recorded by different images, which give a different character, style of actions of a person, and algorithms of his life. In practice, this manifests in the so-called duality of any person. For example: Mr. N is quiet and harmless at work, but at home he is active, loud, and despotic; a different Mr. N uses his intellect at work as an analyst, and during the time of rest chooses physical exercise and, as they say, "does not use his head at all;" some other Mr. N behaves as a leader in communication with several people, but when he is with only one other person—he completely subjects to another's will; and so on and so forth. There are many examples, but the main thing here is that every person (a man or a woman) is "dual" by nature, double-natured. This "duality" is necessary because one cannot be passively recovering while simultaneously actively spending himself.

Existence of separation of individual (subtype) programs into "Active" and "Passive," in our view, can be explained by the fact that sometimes twins have different characters, habits, functionality, qualities of personality. Without getting into details, we will just say that according to our observations, in some cases twins can share a single subtype program; meaning that one lives according to the "Active" part of the program, and the other on the basis of the "Passive." We do not know yet why this happens, more research needs to be done on this, but this fact is the reason twins differ in their personal qualities, characters, algorithms and so on, even though both twins belong to the same subtype.

3.9 What are "six factors of a human?"

In the Catalog of Human Population, which is compiled on the basis of decryption of the ancient monument Shan Hai Jing, all information about a person consists of a general description of an individual and a description of his qualities and functioning on six factors.

Classification by six factors, which are fundamentally different from each other, is a system created by researcher-sinologist Andrey Davydov to describe functioning of human subtypes during compilation of the Catalog of Human Population. Since at a certain point he was faced with a problem: how to structure information obtained from the ancient source about a particular subtype in descriptions of psychophysiological diapason of a subtype? Data obtained from the ancient text contained very diverse information about a human: how representatives of a particular human subtype think, data about their nutritional or environmental diapason, sexual peculiarities, and so on.

In fields that are traditionally related to a human (for example, in medicine) no systems were found, according to which to would be possible to distribute information obtained from Shan Hai Jing. It was necessary to create a classification, which would encompass human life activity in general. To solve this problem, Andrey Davydov identified six factors: intellectual, physical, nutritional, emotional, sexual, and environmental.

And, as it is easy to notice, intellectual factor is only a part of human psyche, and not the most important one. Not to appear unfounded, we will provide an example. If there is certain influence on the physical factor of a human (for example, induction of a painful shock), or influence on the nutritional factor (for example, leaving a person without food or drink for a long time), or influence in a way that a person begins to experience a strong sexual arousal or a strong emotional stress—then, his intellectual factor will turn off, and he will function on the basis of instincts, reflexes, automated processes in his organism. Even a severe headache, as one of the states of the human physical factor, can partially disable the intellect, meaning that a person's thought process will be quite unproductive. Therefore, it was concluded that the intellectual factor is only one of six factors of human psychophysiology.

3.10 Why six factors, and not more?

Number of factors to describe primary programs could not be expanded, supplemented by some other factor. Despite a common misconception that at the first level of development *Homo sapiens* has so-called "spirituality"—information from the ancient text did not confirm this. The appearance of the factor of "spirituality" in humans is a separate topic.

3.11 What is the difference between a woman's and a man's natural program within one subtype?

Men and women within one human subtype differ from each other in the same way as females and males within one subspecies of, for example, animals. They have identical personal characteristics, psychophysiological algorithms, lifestyle, and functionality. Some differences might be present in the sexual factor; for example, in attitudes towards offspring. In nature, females of some species of animals care for their offspring without participation of males; females make males leave their offspring; not only males, but also females of some species do not engage in caring for their offspring at all; or, for example, a male and a female of some types of birds engage in upbringing of their nestlings together and distribute parenting duties equally. However, as mentioned above, these differences are not fundamental.

UNIT 4

NATURAL MANIPULATION MODES OF A HUMAN

4.1 Psychophysiological state and behavior of *Homo sapiens* potentially is 100% manageable; the discovery of natural human manipulation modes.

In addition to detailed descriptions of 293 types of structures of human psychophysiology, Andrey Davydov (researcher-sinologist and author of scientific discovery called the Catalog of Human Population) found descriptions of natural human manipulation modes for each subtype of *Homo sapiens* in the ancient monument Shan Hai Jing. It turned out that a human is subject to operator's influence in relation to himself, as well as to external influence in mediated modes: optimum, balance, suppression, stimulation.

Each person (subtype) has three manipulation modes: suppressing, balancing, and stimulating. However, information about a person's individual (subtype) program, i.e. "optimum mode" is also used to control him.

By nature manipulation modes are built into the unconscious of a human in order to enable him to effectively regulate his psychophysiological state and his behavior. However, as it turned out, if a person knows manipulation modes of a particular individual, and applies this knowledge from the outside, then the individual can be 100% controllable.

The power of this tool for managing a human and its effectiveness are due to that this is a purely natural mechanism. Manipulation modes were not invented, developed by anyone, including by the author of this scientific discovery. Reactions of any person to personal (subtype) manipulation modes occur at the level of unconscious reflexes and therefore no human can resist their application.

Moreover, without knowledge of your own manipulation modes it is impossible to even notice that you are being manipulated. At the same time, when a person knows his manipulation modes, for some time he still continues to react to them. And, this occurs until a certain stage (until which stage specifically is a separate topic).

Representatives of different human subtypes have different manipulation modes. Therefore, some manipulative technique works on one person, but does not work on another.

Also, using the language of science, natural manipulation modes allow to regulate all life activity of a human, but possible metamorphosis from regulation do not change his belongingness to a particular subtype structure of *Homo sapiens*. In other words, it is possible to make an individual feel in a certain way or carry out some actions (any actions and for as long as desired), but, figuratively speaking, a manipulator will not be able to turn a deer into a horse, a rhinoceros into a hippopotamus, a pine into a spruce, and a chamomile into a rose.

As worldwide practice has shown, without knowledge of an individual's subtype manipulation modes from the Catalog of Human Population, it will never be possible to have complete control over a person with a one hundred percent guaranteed result, and our research confirmed this.

4.2 What are suppression, balance and stimulation modes?

Each person has three manipulation modes: suppression mode, balance mode and stimulation mode. Each one of these modes has its own, specific effect on the entire human psychophysiology. These modes are different for different people (meaning, for representatives of different human subtypes).

We will describe each mode very briefly.

Suppression mode, if considered as a mode of self-regulation of a system called “a human,” it rather powerfully relaxes psyche and body of an individual, brings calmness and joy, pleasure, satisfaction, delight. When a person’s suppression mode is transmitted[1], he experiences a state of euphoria stronger than the effect of any narcotics. This completely disorients a person and makes him obey the manipulator, according to the pleasure principle. Right from the start of application of this type of manipulation mode, a person begins to require "a dose" of this kind of pleasure over and over again. And, he will not be able to stop.

When the suppression mode is being transmitted, analysis of the situation and the individual who is transmitting this mode becomes impossible for the subject of manipulation. Any person is absolutely open to this kind of influence and he is defenseless before it because it goes through the unconscious, reflexive sphere, bypasses the intellect, and makes use of those mechanisms, which a person is unable to control.

A pleasant benefit from manipulation when transmitting the suppression mode of a person is that he/she not just falls in love with the manipulator, but literally begins to adore him/her. On this basis, the subject of manipulation is ready to fulfill any request of the manipulator.

The balance mode is called balance because it balances, stabilizes the entire psychophysiology of a person, who is being manipulated, and causes him to be in a harmonious, comfortable state. Balance mode makes the subject of manipulation feel not only very comfortable, but also be quite non-critical. It makes him perceive the manipulator as a friend, even though it is likely that he is not. The comfortable state, which transmittance of the natural balance mode by the manipulator causes, automatically makes a person trust the manipulator, and this trust is unconscious and is not based on absolutely anything. "Doors of the soul" open up.

The stimulation mode causes the subject of manipulation to feel such a strong irritation that a person, so to speak, not only "runs on walls," but also on the ceiling. This irritation turns a person, who is being manipulated into a real kamikaze, and he is ready to carry out any actions that the manipulator needs. Using this mode, it is possible to make any person do absolutely anything; even something that he would never do in his life, as well as something that, perhaps, will bring him irreparable harm, or something that he will regret for a long time.

It does not matter which race, nationality or gender is a person to whom manipulation modes are applied. In this sense, representatives of the entire human population are absolutely the same. By nature each person has an individual subtype program and individual modes of self-control/control of processes of life activities, and behavior in particular.

Of course, this is general introductory information. More details about manipulation modes, as well as what a person, who is being manipulated feels when these modes applied to him, and which feelings make him obey the manipulator, can be found in the book titled Human Manipulation Modes. Either You Are Manipulating Or You Are Being Manipulated (by Olga Skorbatyuk and Kate Bazilevsky). Here, we will add only that a human is unable to guess, "learn intuitively" or "draw out" natural manipulation modes. Even his/her own. This means that he will not be able to control/manage himself or another person with a predictable, stable and guaranteed result.

[1] The term "transmittance of a manipulation mode (manipulation modes)" means that a manipulator acts out manipulative scenarios (or images of manipulation modes) for the subject of manipulation.

4.3 What are the principle differences between the natural tool for controlling a human, which is called Manipulation Modes and all other manipulation methods?

If someone thinks that there are people who cannot be controlled with a one hundred percent result—this only means that he knows nothing about natural manipulation modes of *Homo sapiens* from the Catalog of Human Population.

Every person is born with a built-in mechanism called "manipulation modes," which allows for correction of his psychophysiological state and behavior. It is impossible to control a person with a guaranteed result without knowledge of his manipulation modes. However, with this kind of information from the Catalog of Human Population, it is possible to make any person think, act, feel, live in a certain way. With the knowledge of natural manipulation modes, it is possible to make any representative (or a group) of human population perform any kind of action.

Moreover, everything that a person was motivated to, he explains to himself and others as solely his decision, a manifestation of his free will, and an absolutely voluntary action on his part. And, indeed, this is the case because motivational mechanisms get triggered on the inside. This is very different from a situation when a person is forced to do something, and he realizes that he has to do that what he was forced to do, while lacking an internal desire to do it.

This is very beneficial for a manipulator. A manipulator, who uses manipulation modes of another person, is always "in the shadow." It is impossible even to suspect that he is manipulating someone. And, even more so, it is impossible to blame him. How can you blame a person, for example, for expressing a thought or wearing a certain piece of clothing or ordering a particular dish at a restaurant? It is difficult to even imagine such a thing.

The manipulator is not just guaranteed to receive that what he seeks from the other person, but he is also completely safe from any kind of dissatisfaction, negative reactions, pretensions on part of the person whom he is manipulating.

It is impossible to hide from such influence in any way; at least because in order to manipulate a human's soul and body by the use of his personal manipulation modes it is not necessary to communicate with him in person. It is possible to transmit manipulation modes of a person by communication via social networks, chat rooms, e-mails, telephone conversations, and even text messages. Even if an individual does not leave his home and does not meet up with anyone—this will not protect him from manipulation using his personal manipulation modes.

A lot of money, which allow one to hire bodyguards, immuring in a castle or "laying low" on board of a personal submarine will no longer protect from invasion of "the holy of holies" of a person—his psyche and body. It is now possible to do anything with physiology of any person without the use of violent methods, but exclusively through psyche.

A high social status will no longer protect either. A person at any social level is completely open to this kind of manipulation. At least because in the life of any person there is always someone with whom he/she is in contact. Without contacts any person goes mad and dies, and this is a well-known fact. However, even if we assume that a person does not allow any contacts with other people, still there are many ways to apply his personal manipulation modes to him.

Absolutely any person can now engage in manipulation of others with a guaranteed result; because now no special education or access to classified information (psychotechnologies, special scientific research, compromising materials, etc.) is required in order to do this.

Detailed information about what exactly should be done to "break open" someone else's "soul," is provided in a very understandable form—in the form of manipulation scenarios. Even a poorly educated person is able to understand such scenario. It is enough to simply be able to read. And, it is not necessary to be a professional actor in order to act out a role from a manipulation scenario.

Therefore, it is no longer necessary to be a professional psychologist, a representative of special services, etc. in order to manipulate another person, and to do it effectively.

4.4 What is a Manipulation Scenario?

Anyone who wishes to manipulate another person and to do it with a guaranteed result no longer needs to think about how to do it. Detailed information from the Catalog of Human Population about what specifically needs to be done in order to manipulate someone is available in the form of manipulation scenarios.

A scenario of any of human manipulation modes (suppression mode, balance mode or stimulation mode) specifies in detail the role that a manipulator must act out for a person, from whom he wants to get something, or whom he wants to make behave in a certain way.

Manipulation scenario is a description of the role, which a manipulator must act out.

A detailed manipulation scenario contains a general description of the role and a very detailed description on six factors: intellectual, physical, nutritional, emotional, sexual and environmental.

Any one of detailed manipulation scenarios (of suppressing, balancing or stimulating mode) describes the following:

- ❖ How and what to talk about with the subject of manipulation; what topics to bring up; which values and thoughts to declare, while presenting them as your own; which areas of culture and scientific knowledge to demonstrate interest in; etc.
- ❖ How a manipulator should look when meeting with the subject of manipulation: which clothing styles, set of colors, hairstyles, accessories, make-up, perfume, etc. to use.
- ❖ How to behave if you are planning to have meals together. Meaning, what specific dishes and drinks to order at a restaurant or to cook, to cook or not to cook at home, what should the portions be like, etc.
- ❖ What specific emotional reactions to demonstrate (meaning, to act out) to a person, to whom you are transmitting his manipulation mode, and in which form to express declared emotions.
- ❖ How to behave during sexual contact with the subject of manipulation. Meaning: what types of sex and sexual positions to suggest, what sexual reactions and sexual preferences to demonstrate, while presenting them as your own, and so on. Also, you will learn what specifically to declare in regard to family, children, and parents, while presenting this attitude as your own.
- ❖ A scenario also describes in detail who to present yourself as on the social plane (profession, hobbies, financial status, etc.); how to decorate the interior of the venue where you are meeting; what gifts to give or offer; what preferences to express in terms of the geographical zone, climate, housing, interior, leisure, entertainment,

etc.; what specific behavioral reactions and algorithms of relations with other people to present as your own.

- And so on.

It is very easy to use such a detailed scenario; for any person. All that is necessary is to read the text and to act out a role from the scenario every time you communicate with a person, whom you want to influence. That is—not to act as you want or think you should act with the subject of manipulation, but to act strictly according to the scenario.

4.5 Is it possible to use images, by which a person's manipulation modes are recorded as a tool to manipulate him instead of a manipulation scenario?

When manipulating another person, instead of manipulation scenarios, as this person's decrypted natural manipulation modes from Shan Hai Jing, it is possible to use images, by which his/her manipulation modes are recorded. This will be a bit more complicated than acting out a role according to a scenario. Since in the case of a scenario, all that a manipulator needs to do is simply read the scenario and act out a role that it describes, while images will have to be studied on one's own.

However, it is quite possible to do if the desire to do it is present. All that you will need to do is to study encyclopedias, reference books on the topic of qualities and lifestyle of a natural object, which, as an image, is present in the manipulation mode of a person you are interested in. It is also desirable to have hieroglyphics from Shan Hai Jing (meaning, hieroglyphs, by which images are recorded) and a certain dictionary[1]. After studying images, it is possible to act out all found qualities and characteristics of images for a person, whom you decided to influence.

Images of manipulation modes can also be learned from the Catalog of Human Population. They can be used for manipulation purposes on their own or together with a manipulation scenario in order to better understand the role that needs to be acted out for a subject of manipulation.

[1] When working with hieroglyphics, by with which images of human programs and manipulation modes are recorded in Shan Hai Jing, it is necessary to have the Great Chinese-Russian Dictionary edited by I. M. Oshanin (or a high-quality translation of this dictionary into your native language). Use of all other dictionaries for this purpose is absolutely unacceptable.

4.6 Can I fully control psychophysiology and behavior of a person using partial knowledge and application of his/her natural manipulation modes?

In order to make a person feel and behave in a certain way for some time, or in order to solve some local problem—it is possible to transmit just one of personal manipulation modes of that person.

As for which mode should be used—this depends on the situation. For example, if you decided to make someone fall in love with you or seduce, then the suppression mode should be used. And, if you need to, for example, quickly get rid of presence of some (unpleasant, dangerous, etc.) person, then this person's stimulation mode should be transmitted.

However, in order to make any person one hundred percent controllable and to manipulate him/her with a one hundred percent result—it is necessary to know all four parts of his/her subtype structure: suppressing, balancing and stimulating modes, as well as his/her individual program. And, of course, it is necessary to use this knowledge on practice by acting out all four manipulation scenarios.

In other words, it is possible to influence another person without having complete information about all of his manipulation modes, but the only way to make another person submit fully and gain full control over his behavior can be compared to a car with four wheels, which you will not be able to drive in the absence of even one wheel.

4.7 Why should I know dates of my manipulation modes?

While studying the Catalog of Mountains and Seas (Shan Hai Jing), Andrey Davydov, who discovered in it the Catalog of Human Population, found out that each one of the three manipulation modes of a person is an individual program of some subtype of *Homo sapiens*. Therefore, it turned out that people within the same subtype are not just blood relatives (blood because we suppose that blood might be related to the "software" of a human), but it is also possible to say that they have very close kinship ties with other three subtypes (meaning, carriers of suppression, balance and stimulation modes of an individual).

However, we will not go into the description of this rather complicated natural mechanism—the influence that people have on each other in time and space; we will just say that when a person knows the dates of his manipulation modes, he can find ideal partners for any type of relationship, any kind of activities and any kind of influence on his own psyche and physiology[1].

[1] More on this can be found in the book by A. Davydov and O. Skorbatyuk, which is titled Ideology Of Religion: Scientific Proof Of Existence Of "God"—The Catalog Of Human Population, as, in essence, this book is not about religion, but about the Catalog. Since the Catalog of Human Population (despite the fact that it can become an ideal basis, foundation for any of the existing religions) originally has nothing to do and cannot have anything to do with religion because it is the source of knowledge, and knowledge and faith are completely different things.

4.8 Why should I know manipulation modes of others, and why should I manipulate someone?

Regardless of which social class people belong to, whether they like it or not, they are connected with other people. A lion's share of processes, in which a person participates, either way is contact with another person.

Since any individual normally lives among other people, the problem of knowledge and understanding of another person is everyone's urgent need. However, to date there is no source (besides the Catalog of Human Souls) in civilization, from which a specific John could learn how to communicate with a specific Tom or Mary.

And, not just communicate in some way, but to do it in that exact manner that John needs; and, not just get "whatever" from this communication, but strictly that what he personally needs. In an attempt to achieve this, John uses one method after another (as usual, designed "one for all") until all methods are tried out, and something finally works at least to some extent. However, sometimes it is too late—the relationship is ruined, the contact is broken, and goals in regard to the opponent are not achieved.

We think that the root of this problem is fundamental difference of internal characteristics of one person versus another, despite the external similarity, and sometimes even kinship. Since it is impossible to distinguish subtypes of *Homo sapiens* based on appearance—it is much easier to tell a rabbit from a donkey, despite the fact that both have long ears. And, that might have very unpleasant consequences. Since in nature when one tries to interact with a wild wolf as with a dog, with an eagle as with a dove, and with a venomous snake as with a worm—the consequences are always fatal.

However, it is the same with people. Not knowing the person, with whom you come in contact, it is impossible to know what to expect from him/her and how to interact. Of course, unlike it is with animals, by interacting with another person inappropriately, in the physical sense you will not get bitten, scratched or torn into pieces. (Although sometimes this happens.) However, in psychological or, for example, financial sense—biting, scratching and tearing apart is not at all uncommon among people. Unfortunately, this type of "communication" occurs very, very often. As a result, after communicating with other people, a person very often has to "lick his wounds."

However, in this respect the situation has changed cardinally due to the scientific discovery of the Catalog of Human Population, which was made by Andrey Davydov. Now in order to make any kind of contact it is not necessary to gather information about a person from third parties or from other sources of information, there is no longer a need to use the well-known "scientific method of trial and error," and so on.

Any kind of contact (business, friendly, intimate, and so on) with any person is now done quite easily using other methods. In order to know what to expect, and what not to expect from a person, all that is needed is information about the individual program and manipulation modes of that person.

Application of this kind of information guarantees establishment of any type of required communicative contact and with any required results. Any interaction can easily be modeled using information about another person from the Catalog of Human Population. And, as a result, any person can get that what he planned, that what he wants and needs from another person.

The success of use of this technology is not affected by gender of the subject of interest, or by his/her age, educational level or professional affiliation, financial situation or belongingness to any social group. Information about another person from the Catalog of Human Souls allows anyone to have an ability to influence, make adjustments to behavior, plans, and objectives in relation to that person.

Consider any of the problems, for example, within tandems "husband-wife" ("lovers"), "parent-child" that we can safely state are common in 99.9 percent out of 100. The root cause of any one of them is the same: lack of knowledge of another person, inability to properly communicate with him, and inability to influence him. These are, so to speak, the "three pillars," upon which in this civilization all communication between one person and another stands.

Only because of these "three pillars" parents at some point lose understanding of behavior of their children and control over their actions, and children are not able to "get through to parents," to explain their needs, aspirations, life stance to them. Also, this is the only reason why at some stage men and women go in different directions and live separately—they are simply unable to neither find a person, with whom they would be happy, nor find a common language with a person, who for some reason they selected for family life or for other type of relationship. And so, they live apart, despite that both women and men, regardless of how old they are, do not want to be alone. They want to feel someone's shoulder close to theirs, they want to be needed, loved, and wanted.

To us, all problems listed above seem unworthy of wasting one's breath on, as they are easily solvable by using information from the Catalog of Human Population: the so-called search for "the other half," and mutual understanding, and love between parents and children or between a husband and a wife, and how to find a common language with another person, and how to be liked, and how to influence, and so on.

So-called "ordinary people" can dream as much as they want that if he or she is not involved in something, from their point of view, serious and dangerous, they should not worry about their safety or preservation of their well-being. From our point of view, this is a very big delusion.

We live in a civilization where each one of us, figuratively speaking, lives his whole life in the area of active combat operations. Regardless of whether he knows about it or not, notices or does not notice it. It is just that the scale of these combat operations (without quotes) is different for different people: for some it is the process of conquest of economic and political influence over entire regions, and for some it is fighting over, figuratively speaking, a rusty frying pan.

It so happened that in this society that in any type of relationship between people—actions against each other cannot be classified in any other way except by the word 'war': in business areas, in so-called "friendship," and even in families. And, in reality it is very difficult not to notice this fact. As a rule, it is not that people do not notice it, but they just do not want to think about it, hoping that their games of peaceful life with each other, their false assurances of friendship and love will protect them from an attack. No, they will not protect.

And, it should be noted that consequences of participation in such combat operations are just the same for a major politician as they are for an ordinary, simple citizen—pain, psychological, physical, financial and other damages, etc., up to physical destruction. War is war. Hence, the joke: "How do you feel, Joe?" – "Thank you, not well at all!" Life at war is difficult and it is even more difficult to survive. However, hiding in fantasies is not just pointless, but extremely dangerous.

The whole question is not in ambition since we faced the fact that ambitions of so-called "ordinary people" sometimes can be truly exorbitant, but in that what a person is capable of (himself, him personally); in what information he has, what he knows how to do, how he does it and does he do anything at all. Since sometimes people behave quite strangely; for example: they complain, get offended, upset, when someone reaches any success or surpasses them in their personal qualities in reality. These people are so used to being great and mighty in their fantasies that any encounter with real facts, which confirm that they are really not, make them very depressed and they feel hate towards other people. From our perspective, sometimes this reminds of a situation with a dog, which lazily wiggles its tail all day in anticipation of getting fed and petted, or sleeps, pretending that it protects the house during the breaks, and with the return of the master after a hard day of work is puzzled and upset after getting only bones off of a piece of meat.

Yes, usually no one physically destroys so-called “ordinary people” for their mistakes in communication, for misunderstanding and miscalculation of a situation, as it happens with, for example, politicians, officials, business leaders, etc. However, no one is immune to psychological destruction by another person or a group of people, which, as it is known, is followed by either physical destruction or, more commonly, physical self-destruction because psyche is primary. Usually this does not get taken into account right away, and when something happens it is usually too late to do anything about it—you, so to speak, got eaten. This society is a jungle, Ngorongoro, where predators hide behind masks of innocent white sheep.

4.9 Why natural human manipulation modes are the only alternative to violence? And, what about manipulation ethics?

From our point of view, manipulation modes of a human were created by nature as the only way of effective, constructive, non-deforming, and secure communication.

Based on that application of natural manipulation modes in communication with any person always exceptionally favorable affects his psychophysiology (and not on the basis of the meanings of what manipulation is usually considered)—these modes were named manipulation modes. Contrary to the belief, which is popular both in psychology and in everyday life that manipulation is always a process with a negative connotation, we consider manipulation based on other meanings from dictionaries. We consider manipulation to be any purposeful action with a subject (refer to a dictionary).

From our perspective, communicative impact that leads to actualization of quite specific motivational states in a subject, and makes him behave in a certain way, should not necessarily be beneficial to the manipulator and not beneficial to the subject of manipulation.

From our point of view, this directly depends on ethical and moral qualities of a person applying manipulation modes towards another person. Just like a knife cannot be regarded only as a murder weapon. A knife as a tool can also be used for cooking, as a scalpel in the hands of a surgeon, and so forth.

Of course, being very powerful, this purely natural mechanism can be used by some to have the other person submit, in order to use him for some purposes. Indeed, when personal manipulation modes are applied to a human, he or she forms a complete willingness to do anything for the manipulator. However, we consider use of natural manipulation modes with immoral goals—inhumane, immoral, and contrary to nature.

Whether someone likes the fact that any person has a natural mechanism called "manipulation modes" or not—it exists anyway. And, the important thing is that this is how nature arranged it. Therefore, in the absence of the Catalog of Human Population it is, of course, possible to continue to treat *Homo sapiens* like an animal, while discussing humanism, goodness, morality and other things unrelated to this process. This is widely practiced in this society. Consider a method that in the case of animals is called training, and in the case of a human for some reason is referred to by different words: upbringing, re-education, etc. Zoopsychologists are very well familiar with this method as the "method of positive and negative reinforcement;" animals are trained using with method.

However, unlike subtype programs of animals, in which only instincts and primitive reflexes are recorded, human programs consist of much more than that. From our point of view, Pavlov's (a Russian researcher, who was involved in elaboration of certain reflexes in dogs) method that has become popular worldwide should not be practically the only available method of influence on a human being in the XXI century. Influence of a human must be based on more complex principles than induction of salivation.

Our studies have shown that human manipulation modes are a subtype's peculiar language. If a person is not spoken to in this "language," then he practically does not hear his interlocutor, does not want to understand him, does not accept or simply rejects him. Our research of Shan Hai Jing showed a firm link between this "subtype language" and individual subtype program, meaning—human soul. Therefore, knowledge and use of natural "language of the soul," meaning manipulation modes of a person, allows to not degrade him by treating him like an animal. And, this makes the use of violence against a person in order to motivate him to some actions or psychophysiological states unnecessary. A human does not need to be forced, it is not necessary to demand something from him—all that is necessary is to communicate with him using his "subtype language," which is the language of his natural program and manipulation modes.

UNIT 5

IMAGES FROM 山海經 (SHAN HAI JING) ARE THE LANGUAGE OF HUMAN "SOFTWARE," PATTERN, BLUEPRINT, BASED ON WHICH NATURE CREATES *HOMO SAPIENS*.

5.1 What do researchers of 山海經 (Shan Hai Jing) decrypt in order to obtain information about a particular person from this ancient source?

As it was found by researcher-sinologist Andrey Davydov, the author of the scientific discovery of the Catalog of Human Population, any kind of information about any representatives of the biological type *Homo sapiens*, who lived in the past, live today, or will live in the future is encrypted in archaic images[1] of one of the most ancient monuments 山海經 (Shan Hai Jing, Catalog of Mountains and Seas). The researcher also found that in order to decrypt the information from this source, it is necessary to decode hieroglyphs using special technology. Andrey Davydov created the technology to decrypt images of human programs and manipulation modes from Shan Hai Jing in the early 80s of the XX century.

Some of the hieroglyphs in Shan Hai Jing (Book of Mountains and Seas) are so ancient that they are hard to find in modern dictionaries. This ancient source, which, as Andrey Davydov discovered, preserved knowledge of functioning of a human and his psyche for many thousands of years[2], contains a huge number of hieroglyphs. For example, there are over fifteen thousand hieroglyphs just in the first section of the text titled the Catalog of the Mountains, which consists of five parts out of existing eighteen. And, behind each one of these hieroglyphs (of course, excluding conjunctions, prepositions, and parts of speech that do not carry semantic meanings) is some archaic image, or, using the language of psychology—an archetype.

However, when discussing archetypes, the following must be noted right away. The word 'archetype', introduced to psychology by C. G. Jung[3], refers to so-called "archetypes of the collective unconscious," i.e., collective archetypes, while in the case of archetypes contained in the ancient monument 山海經 (Catalog of Mountains and Seas) we are discussing individual archetypes[4]. Archetypes in Jungian analytical psychology, along with instincts, are inborn psychical structures located in the depths of "the collective unconscious," whereas archetypes listed in the Catalog of Mountains and Seas are a peculiar matrix, an imprint[5], recorded inside an individual, in his personal, individual unconscious. And, these archetypes do not in any way relate to "the collective unconscious."

In understanding of followers of Jungian direction in psychology, specificity of archetypes as psychical contents is that they carry properties of all humanity as something whole. However, archetypes, as images of human's natural programs and manipulation modes from the ancient Chinese monument 山海經 (Shan Hai Jing), which is being decrypted by developers of the Catalog of Human Population, carry properties of only one of 293 subtypes of *Homo sapiens*, not of all humanity.

In conclusion, it should be noted that an archetype in that sense of the word, in which it is present in the culture of this civilization in general and in psychology in particular[6], significantly differs in essence from the archetypes, by which human "software" is recorded, and by which, as researcher of 山海經 (Shan Hai Jing) Andrey Davydov found, all information in psyche of *Homo sapiens* is recorded.

All this combined differs very significantly from the existing understanding and interpretations of the word 'archetype', including from Jungian analytical psychology. Therefore, despite that in both cases the word 'archetype' means preimage (prototype), when it comes to the language of human "software" recorded in the ancient monument 山海經 (Shan Hai Jing), we prefer to use the word 'image' rather than 'archetype' in order to avoid misinterpretation. We use the word 'archetype' mainly in scientific articles and in definitions of Non-traditional Psychoanalysis (a new scientific direction in psychology, which is based solely on decryption of Shan Hai Jing) in order to show that we are discussing a preimage, a prototype, as the fundamental principle of existence and functioning of *Homo sapiens*.

Since images in the ancient monument 山海經 (Catalog of Mountains and Seas), which turned out to be the source of detailed descriptions of psychophysiological structure and functioning of 293 subtypes of biological type *Homo sapiens*, are archetypes as a fundamental principle, based on which any specific individual exists—these images (archetypes) are being decrypted by the founders of Non-traditional Psychoanalysis—Andrey Davydov and Olga Skorbatyuk—in order to create descriptions of subtype structures of *Homo sapiens*, while compiling the Catalog of Human Population.

[1] Archaic, archaistic (from the ancient Greek ἀρχαῖος – "ancient") means original, initial, from the beginning; C. G. Jung defined it as "ancient character of psychical contents and functions." It is considered that the quality of an image is archaic when the image has doubtless mythological parallels.

[2] Exact dating of the monument 山海經 (Shan Hai Jing, Catalog of Mountains and Seas) is still not determined, but some researchers date this ancient monument XXIII century BC.

[3] Archetype (from the Greek arche – the beginning, and typos – an image) means preimage, an original image, an idea. In analytical psychology, founded by Carl Gustav Jung, an archetype means universal inborn psychical structures that make up contents of the collective unconscious. "That what we call instincts is a physiological motivation and is perceived by the senses. However, at the same time instincts manifest themselves in fantasies and often reveal their presence only by symbolic images. I called these manifestations—archetypes." (Jung, C. G. (1991). A. M. Rutkevich (Ed.). *Arkhetip i simvol* [Archetype And Symbol]. Moscow: Renessans.)

[4] Davydov, A., & Skorbatyuk, O. (2014). K. Bazilevsky (Ed.). Anonymous (Trans.). *Archetypal Pattern. Fundamentals of Non-Traditional Psychoanalysis*: *Vol. 1. From Carl Gustav Jung's Archetypes of the Collective Unconscious to Individual Archetypal Patterns*. San Diego, CA: HPA Press.

[5] ἀρχή "the beginning" + τύπος "impact, footprint, imprint", ἀρχέτυπον — preimage, ἀρχέτυπος "initial, original."

[6] For example, it is considered that archetypes are inborn conditions of intuition, i.e., those components of all experience, primeval forms of comprehension of the outside world, timeless schemes or bases, according to which thoughts and feelings of all humanity generate, and which originally include all the richness of mythological themes or collective sediment of historical past that is stored in people's memory and makes up something common, immanently inherent to the human race and so on.

5.2 How does the technology of obtaining information about the unconscious sphere of an individual, practiced in Non-traditional Psychoanalysis that is based on decryption of images (archetypes) from 山海經 (Catalog of Mountains and Seas), differ from other psychodiagnostic methods?

Technology of decryption of archaic images, by which information in the psyche of *Homo sapiens* is recorded, was created by researcher-sinologist Andrey Davydov (the author of scientific discovery of the Catalog of Human Population) in the early 80s of the XX century.

This technology finally made it possible to obtain objective, complete and reliable information from the unconscious sphere of *Homo sapiens*. It is very different from all preexisting and currently used psychodiagnostic methods and technologies of penetration into the unconscious sphere of a person in results that a researcher obtains. Since results of decryption of 山海經 (Shan Hai Jing) have nothing to do with manifestations of anyone's subjectivity, fantasies, assumptions. The technology of obtaining information from the unconscious sphere of *Homo sapiens*, which was created by A. Davydov, completely excludes any inclusion of subjective interpretations, and, in essence, is a dry translation from the language of archaic images from the ancient monument 山海經 (Catalog of Mountains and Seas) into the language of modern psychological description of an individual.

This technology provides objective and very detailed information about any individual of interest: what are the qualities and properties of his personality, his life algorithms, functioning, abilities, desires and goals, hidden motivational springs, etc. This became possible only because archaic images listed in 山海經 (Catalog of Mountains and Seas) are that foundation, cornerstone, on the basis of which *Homo sapiens* was created and functions.

By finding out with what images (individual archetypes) the natural program and manipulation modes of a person are recorded in the ancient source 山海經 (Shan Hai Jing) and by decrypting hieroglyphics, by which these images are recorded, developers of the Catalog of Human Population—A. Davydov and O. Skorbatyuk—have a possibility to learn how any person of interest is arranged and functions.

Moreover, archaic images from this ancient source, which turned out to be the Catalog of Human Population, reveal not only how exactly a person will present himself to others, but also that what is behind his regular self-presentation: hidden motives, desires and objectives, about which an individual usually does not reveal anything to anyone, etc. These true motives, desires and goals are sometimes hidden by an individual even from his own self. And, more often he is not conscious of them at all.

In essence, decryption of hieroglyphs from the ancient source 山海經 (Shan Hai Jing) provides access to contents of that part of human psyche, about which not only other people, but the individual himself has no idea. An individual is unable to control this kind of psychical manifestations and is unable to correct them—he simply acts on their basis, automatically. And, no *Homo sapiens* is able to act otherwise because these psychical manifestations, which are recorded by images in his unconscious, is "software" that is built into him from birth, which he, as a bio-robot, is unable to bypass (at least, without damaging his psychophysiology).

The true causes of his actions or psychophysiological states that are implanted specifically in archaic images (archetypes), by which his natural program and manipulation modes are recorded, an individual is unable to explain, for example, to a psychologist. And, based on external manifestations of functioning of a particular image (archetype)—it is not possible in principle to uncover and understand to what image of the program of the unconscious sphere of an individual these manifestations relate.

Probably, this is the root cause of problems of any one of psychodiagnostic methods that exist today in psychology, which do not use information from the Catalog of Human Population. All these problems are easily solvable with the help of decryption of hieroglyphics from the ancient source 山海經 (Catalog of Mountains and Seas), which turned out to be the Catalog that describes subtypes of biological type *Homo sapiens*.

It is important to note that the technology of decryption of images of human's natural programs and manipulation modes that are recorded in 山海經 (Shan Hai Jing, Catalog of Mountains and Seas) opens up access to the unconscious sphere of any one of over 7 billion *Homo sapiens* living on planet Earth. Since physiological properties of a particular human subtype, which are implanted in him from birth through images (archetypes) in his unconscious, are not changed by that an individual belongs to a particular race, nationality, and do not depend on place of his birth and residence, and similar factors, which are only insignificant correctors that do not change the essence of archetypal images implanted in his psyche.

5.3 What is the carrier of imagery information, by which all qualities, functions, abilities of a person, in short—his whole life is recorded?

As it was found by Andrey Davydov, the author of scientific discovery of the Catalog of Human Population, psyche is the carrier (or, using computer terms—"a hard drive"), where all information about a particular representative of the biological type *Homo sapiens* is recorded from birth.

Of course, scientists have long supposed that psyche is the carrier of some important information about a human. In the last century C. G. Jung (a student of the founder of psychoanalysis—Sigmund Freud) stated: "Immensely ancient psychical beginning forms the basis of our mind, just as the structure of our body goes back to the general anatomical structure of mammals."[1] However, all those who justly considered that psyche is primary and lies at the basis of all life activity of an individual did not state that it is the space where such information about *Homo sapiens* as his natural subtype program and personal manipulation modes are recorded from birth. In reality, it turned out that no human child is born "a tabula rasa" ("a blank slate"), i.e., empty, without a program of life activity implanted in his psyche.

Another confirmation that the carrier of all information about a person is his psyche is a huge amount of factual material in psychiatric clinics. There are documented cases in psychiatry when it was noted that a patient showed a significant change in his behavior cases when the primary "record" was erased due to some reasons, but a person remained alive. Lack of mental functioning allowed patients to look like humans, but nothing human was observed in their behavior anymore.

If we consider psyche from the point of view of its already long-held division in psychology into the part available for conscious perception by an individual ("consciousness") and the part that is not possible to consciously realize ("unconscious"), then it is clear that human "software" in the form of natural subtype programs and manipulation modes is implanted in the unconscious. Since, as many years of research have shown, natural subtype program of an individual and his reactions to his personal natural manipulation modes take place at the level of instincts, unconditioned reflexes, and without use of special methods can be consciously realized only in part, only by a few percent.

[1] Jung, C. G. (1991). A. M. Rutkevich (Ed.). *Arkhetip i simvol* [Archetype And Symbol]. Moscow: Renessans.

5.4 In what language is information about natural "software" of *Homo sapiens* recorded?

As it was found by researcher-sinologist Andrey Davydov (the author of scientific discovery of the Catalog of Human Population), recording of psychophysiological algorithm of *Homo sapiens*, on the basis of which human activity builds and happens on six factors (intellectual, physical, nutritional, emotional, sexual and environmental), occurs in the psyche; this recording is made by archetypal images, and the order of recording of subtype characteristics depends on phenological cycles[1].

In the mid-70s of XX century Andrey Davydov began to study the mysterious ancient Chinese monument 山海經 (Shan Hai Jing, Catalog of Mountains and Seas) and found that this source is the richest collection of archaic images. Other ancient sources, which he researched, also contain images, but the Catalog of Mountains and Seas is an undeniable leader by concentration of images per page of text, and their number exceeds all limits of imagination.

Despite that the abundance of images in any source is considered by many researchers as a primitive ancient mythological perception of the ancients about themselves and about the world, Andrey Davydov decided to look at it differently. He was not of the opinion that the ancient people were more foolish and naïve than modern, and thought that it was unlikely that they would engage in such an unserious and useless act as creation and recording images of fantastic animals, which have never existed even as myths, in a written source. After all, in order to do this, they would have to put in a lot of work and mental efforts, whereas in order to get food our ancestors had to work hard day and night under harsh conditions.

Researcher-sinologist Andrey Davydov suggested that archaic images from 山海經 (Shan Hai Jing, Catalog of Mountains and Seas) contain not only some very important information, but contain information about a human and his arrangement; and, first of all, about the most mysterious, the most unknown constituent of *Homo sapiens*—his psyche, soul. Being familiar with research of psychologists, who argue that the archaic layer of the psyche is the unconscious, where "live" archetypes as preimages (prototypes) of any kind of human behavior and state, the researcher attempted to uncover the mechanism of "input" of ancient archetypal information into the unconscious.

Since such a record contains all information about an individual, including the smallest details, according to calculations of the researcher—it must be sufficiently volumetric. Therefore, there was a peculiar kind of archiving, that is— compression of information to an acceptable volume and form. In the works of theologians, philosophers, philologists and other scholars, there were views expressed about an image as a certain formula, by which information is transmitted without change, complete, and the creation of any image—one of the highest achievements of a human, as an impress of his understanding of principles of creation of the world. However, upon receipt of an image (or images) it is necessary to decrypt it into speech, text, dance, into action, which by the volume of decrypted material exceeds the graphical volume of the image itself.

Analogous to such a process are electronic data carriers[2], where the volume of information is much greater than the volume of the carrier. Since an image as a hologram (meaning—a volumetric image), carries information compendiously, multidimentionally in space and outside of time, Andrey Davydov suggested that it is an image that is most suitable for such a recording.

A child's consciousness (soul, unconscious) records a dynamic picture of the surrounding world through mythological images, as a universal variant for understanding, and as the most capacious process of recording in mythical images, where the given element is holographically volumetric for consideration when it is necessary to turn to, as a kind of cheatsheet, and "return to himself," natural. However, if each person has the possibility to "return" to himself, it is difficult to suggest that the entire population of the Earth has the same recording, equal to a recording of one person.

Therefore, our remote ancestors or creators could be so careful that they recorded fundamentally important points, saved them in some encyclopedia, directory, catalog. So that if someone forgot something, or is simply interested to know other people's "recordings," he would "turn the pages," refresh his memory, and satisfy his curiosity. To all appearances, such an archive could be mythological, displaying all metamorphoses of a human and nature on the whole. From this, Andrey Davydov suggested that the ancient monument 山海經 (Shan Hai Jing) is that catalog—the catalog, in which images of all models of psyche of *Homo sapiens* that exist in nature are recorded[3].

The process of exploration of this question took over decade, included a lot of research in various fields of scientific knowledge and experiments. In the end, Andrey Davydov's suggestions were fully confirmed: the ancient

Chinese monument 山海經 (Catalog of Mountains and Seas) indeed turned out to be the description of structure of psyche of *Homo sapiens*, and strange, sometimes fantastic images from Shan Hai Jing—that language, by which all information about an individual, as a representative of one of 293 human subtypes, is recorded in his unconscious.

The ancient Chinese monument 山海經 (Shan Hai Jing, Catalog of Mountains and Seas) was qualified as the Catalog of Human Population on the basis that this source is a preserved ordered system that in practice allows to uncover the structure of the unconscious, where the root system of all human motivations and any other human life activity is concentrated. And, the main proof was the fact that if it is known what images from 山海經 (Shan Hai Jing) relate to a person of interest, as a representative of a particular subtype of *Homo sapiens*[4], and there is knowledge of how to decrypt them, then it is possible to get a very detailed description of what psychophysiological properties this individual possesses, to learn how he lives in detail and what true motives lie behind his actions, as well as how his psychophysiological state and behavior can be controlled.

[1] Phenology is a system of knowledge and sets of information about seasonal phenomena of nature, time periods of their occurrences and reasons that define these periods, as well as the science of space-time laws of cyclical changes of natural objects and their complexes, associated with annual movement of the Earth around the Sun. In nature, absolutely all bio-forms (plants, animals, birds, fish, insects, reptiles, etc.) appear during their time periods. Therefore, in order to find the description of a certain person in the Catalog of Human Population, it is necessary to know the day, month and year of his/her birth; and, this has nothing to do with neither astrology, nor numerology.

[2] In scientific monographs (end of XX—beginning of XXI centuries) Andrey Davydov compared this with magnetic carriers, such as a computer diskette or a laser disc.

[3] In the ancient Chinese monument 山海經 (Shan Hai Jing, Catalog of Mountains and Seas) Andrey Davydov found detailed descriptions of 293 archetypal structures of human psyche, which clearly indicated that the biological type "human" is divided into subtypes, each of which has a recording in the unconscious from birth about his own, unique archetypal pattern.

[4] The schedule of appearance of subtype structures of type human is copyrighted property of Andrey Davydov and Olga Skorbatyuk—developers of the Catalog of Human Population.

5.5 What is the role of an image for a human and his psychophysiology from the standpoint of Non-traditional Psychoanalysis, as a new direction in scientific psychology?

Fundamental research in the field of human psychophysiology that was carried out by Andrey Davydov, the author of discovery of the Catalog of Human Population, showed that an image is a pattern, a blueprint, based on which nature creates *Homo sapiens*; an image is the basis of all human life activities.

However, even in the enlightened XXI century, unfortunately, few people know that the basis, the foundation of existence and functioning of any human being is specifically an image. Modern people even managed to make up a fairytale that a human thinks with words, despite that this contradicts logic and already elucidated scientific facts; for example, prior to emergence of any word in the human brain, first there is a representation of an object, phenomenon or action—that is, an image, a picture.

First, a picture emerges in the human brain, then a short "video," and only then a person makes some actions, says some words, experiences some sensations or states. However, this process occurs at a great speed, and people do not manage to track it, but this does not change the essence. A thought, in essence is not related to words. And, it is not difficult to suggest that if *Homo sapiens* thought at the speed of their spoken words, then the speed of their thinking would not be any different from thinking of patients, who suffer from psychiatric illnesses.

An image is the basis of any action or state of a person. However, something else is more important here: according to discoveries made by geneticists, audiovisual "packets" of information, which cells of the human body exchange with one another, contain images[1].

An image works at the cellular level. And, this means that a cell of a human organism executes orders of images unquestionably, without discussion. A human cell, figuratively speaking, is a soldier and executes the will of the image-general without contemplations, discussions, and even more so without arguing.

An image is a very powerful mechanism, it, so to speak, "gets into the holy of holies"—the human soul—and sets its own rules there. Probably it is not without reason that in the ancient treatise Tao Te Ching, the author of which is the sage Lao Tzu, an image is put ahead of everything else. When translation of this treatise is done correctly, it is stated there that Tao (Dao) births an image, which precedes the Primogenitor and so forth.

As Andrey Davydov found, the soul (psyche) of a human is the carrier of archetypal information, which defines the subtype structure of a human and, therefore, completely determines the work of his physiology. And, archetypes are images. In this connection, human psyche and body can be compared to characters from the famous work of Jean-Jacques Rousseau titled Pygmalion—Pygmalion and Galatea; where Pygmalion can be considered a psychophysiological algorithm and Galatea a body, which transforms depending on the creative potential of this algorithm. Pygmalion is psyche, as the carrier of the essence, and Galatea is physiology, space, as a field of action of archetypal essences.

The mythological story of Pygmalion and Galatea was mentioned for a reason. As Andrey Davydov and his colleagues found, if the image of Pygmalion is taken as a principal mechanism of influence on material, including on bio-material, then natural structure of the soul influences human physiology like the analogous mythological character. Form as a whole, as well as individual parts of a human are "conductors" of his psyche, which with time get characteristics that correspond to psychical analogues. The latter, as it was found, are a set of various forms that exist in the natural environment.

Results of many years of research done by Andrey Davydov showed that, if a myth is a psychical value[2], and images are a connection of functional values that represent the entirety of a bio-form (human) and are peculiar formulas of psychophysiology of a human, then it makes sense to turn to 山海經 (Catalog of Mountains and Seas), as the only classified, systematized source that is understandable at any level, including molecular. Research and experiments done by representatives of Non-traditional Psychoanalysis clearly showed[3] that ancient times have left us a wonderful method for determining archetypal human programs in the form of this ancient monument, which turned out to be the Catalog of Human Population.

Of course, this is a very brief summary of what an image is. However, it is easy to notice that results of many years of research of 山海經 (Shan Hai Jing), which took place and is continuing by Andrey Davydov and his colleagues, the author of scientific discovery of the Catalog of Human Population, led to a deeper and fuller understanding of what an image is than how it is presented in psychological and philosophical dictionaries[4] because such definitions do not describe the true essence of an image and its true significance for a human.

[1] "...genetic code and genetic texts of living organisms are represented as image-semantic constructs of soliton type." (Gariaev, P. P. (1994). *Volnovoy genom* [Wave Genome]. Moscow: Obshchestvennaya Polza.)

2 "...myths are first of all psychical phenomena, which express the deep essence of the soul." (Jung, C. G. (1991). A. M. Rutkevich (Ed.). Arkhetip i simvol [Archetype And Symbol]. Moscow: Renessans.)

3 If it is known what images from 山海經 (Shan Hai Jing) relate to a person of interest, as a representative of a particular subtype of *Homo sapiens*, and there is knowledge of how to decrypt them, then it is possible to get a very detailed description of what psychophysiological properties this individual possesses, to learn how he lives in detail and what true motives lie behind his actions, as well as how his psychophysiological state and behavior can be controlled.

4 For example: "An image is a real and ideal form of reflection of objects and phenomena of the material world in the human consciousness; an image on the sensory level of cognition-sensation, perception, notion at the level of thinking-understanding, judgments, conclusions. Material forms of realization of an image are practical actions, language, various symbolic models. By content, an image is objective to that extent, to which it adequately reflects an object." (*Sovetskiy Entsiklopedicheskiy Slovar' (SES)* [Soviet Encyclopedic Dictionary]. (2nd ed.). (1982). Moscow: Sovetskaya Entsiklopediya.)

5.6 What is the relation between objects of natural environment that are listed in Shan Hai Jing (such as mountains, seas, rivers, lakes, creeks, streams, rain, wind, various plants, animals, minerals, soils) and human psyche?

Natural archetypal images recorded in the ancient Chinese monument 山海經 (Shan Hai Jing, Catalog of Mountains and Seas), which was qualified as the Catalog of Human Population, are most directly related to human psyche—they are the language of "software" of *Homo sapiens*. Human psyche is like a peculiar conglomerate of natural analogues or their parts.

Andrey Davydov, the author of scientific discovery of the Catalog of Human Population, managed to find out that from birth every person is a carrier of an individual set of archetypal images. The system of archetypal images holds quintessence of the functional diapason of nature, which provides analogues as existing bio-forms, and a human has an individual archetypal complex that is distributed according to phenological cycles and determines his subtype place.

Modern psychology has already proven that from the moment of birth until the moment of maturation—psyche of each one of us absorbs from the environment everything that might be useful in the future. Parents pass on basic information to children and, being next to them, demonstrate how to use it. However, it is impossible to affect a child's thinking with pedagogical influences, holistic thinking of a child does not get implicated in pedagogical tricks, and so something gets taken in, and something does not. On the other hand, each person after being born in a specific geographical "pot" absorbs the areal of his environment—animals, birds, reptiles, plants—genetic and psychological.

Researcher-sinologist Andrey Davydov found that form as a whole, as well as individual parts of a human are peculiar "conductors," which manifest elements of his psyche. In the time parameters, it makes "conductors" acquire characteristics that correspond to psychical analogs or their segments in nature; in other words—representatives of the surrounding flora and fauna. In a sense, this is confirmed by results of studies presented by Stanislav Grof, where there are described cases of when people consciously reach states of consciousness, in which they self-identify with animals, plants, chimeras. These cases are well-known from ancient shamanic, yogic practices, as well as from modern research.

An image carries in itself elements that are identical to natural. Any natural images (such as a mountain, a river, rocks, minerals, metals, animals, plants and so on), in addition to being universal in visual understanding by people of different linguistic belonging, are also indestructible. They existed, exist and will exist as long as planet Earth exists. Therefore, the language of images will be alive while nature exists on this planet.

Uniqueness of natural images also consists in that processes, which are built into natural magnitudes, by their essence do not change over the centuries—plants, animals, rivers, lakes have the same processes now, as they had millions of years ago. As for artificial images (meaning—images, which were thought-up by people themselves)—not one of them influences and, in principle, cannot influence human psychophysiology in a way that any natural image does; and, they do not live through centuries. All artificial images are like mayflies: once there was King Kong or Tarzan, then Godzilla or Superman, and then Ironman or Shrek. Each one of such images "lives," on average, for several years or maximum a few decades, and then it is forgotten or gets driven out by new inventions of *Homo sapiens*.

None of the images thought-up by people live for centuries, and all the more so billions of years, as it is the case with natural images. This is exactly the reason why, possibly during creation of *Homo sapiens* a natural image was put into the basis of his existence and functioning.

5.7 What is the difference between natural images and artificial images?

There are many images present around us. Images are on every corner, simply everywhere. However, all images, which in this civilization are created by people—are artificial images, whereas information in human psyche is recorded exclusively by natural images.

Natural and artificial images differ very significantly in terms of their influence on human psyche and physiology. To put it briefly, the main difference between natural and artificial images is in that natural images transmitted to a cell of a human organism give termless health, youth, life, while artificial—slowly kill.

An image, as Andrey Davydov (the author of scientific discovery of the Catalog of Human Population in the ancient Chinese monument 山海經 (Shan Hai Jing)) found out, really is the basis of all things, as the ancient Chinese philosopher Lao Tzu made it clear in his treatise titled Tao Te Ching. And, it is quite likely that instead of "In the beginning was the Word" the early Bible stated: "In the beginning was the Image." After all, the Bible was translated from the Greek language, and in Greek philosophy 'eidos' (an image, a prototype, an archetype) does not occupy the last place.

There are preserved stories about ancient priests, who, using a tool called "an image," could manage not only any number of people, but nature itself. Modern "puppet masters" struggle to repeat this. However, without knowledge that priests had, "lords of human souls" are unable to go further than creation of most primitive artificial images.

The secret of priests is very simple. Images can be natural and artificially created. And, images, as the basis of all things, refer to natural images and not in any way to ones thought-up by people.

Also, "puppet masters" probably do not know that every single artificial image has a destructive influence on human psychophysiology, slowly kill him from the inside (and, its creators as well). Also, they probably do not know that in this civilization exists a source that is a collection of a huge number of natural images—this source is 山海經 (Shan Hai Jing).

It would seem that this ancient monument could save "lords of human souls" from failure since natural images, as opposed to artificial that they invent, give real power over a human. However, it is not so. It is not enough to find out these images from 山海經 (Shan Hai Jing)—it is also necessary to have information to which person (which subtype) certain images relate to in the ancient monument, as well as to have information about how to work with these images[1]. Without having all of this information in a complex, the ancient monument 山海經 (Shan Hai Jing) becomes absolutely useless, and unsuccessful experiments lead to enormous damage of psychophysiology of experimenters.

[1] Tracking of transformation of natural forms into archetypal images and identification of patterns that allow uncovering of archetypal complexes that are implanted directly in the human soul are copyrighted.

5.8 "Natural images are etalon-carriers for a human"—what does this mean? And, how are images of a human program or a manipulation mode considered through natural analogues?

During their time many inquisitive minds explored similarities and differences between humans and animals, and have not managed to find the cornerstone that allows to confidently state anything specific about this. In the end, they resorted to concepts of "soul" and "spirit," which, in their view, distinguish a human from an animal. However, for representatives of Non-traditional Psychoanalysis these conclusions seem either too premature and unconvincing or, for example, in regard to "spirit"—frankly erroneous.

Fundamental research in the field of human psychophysiology, carried out by the author of discovery of the Catalog of Human Population—Andrey Davydov and his colleagues, showed that an image is an ancient formula, expressed by a natural bio-form.

Natural images are etalon-carriers as examples of specific qualities present in a certain physical form. And, their functionality can be translated onto functionality of human psychophysiology.

Human psyche as a whole is like a peculiar conglomerate consisting of separate assimilated forms and has the same diapason of actions as a natural analog or a part of an analog. It would be overhastily to assign only the role of food, incidental material, etc. to animals and plants since the development of civilization has a vector directed to consideration of various bio-forms as concomitant, without their consumption by a human.

Since Andrey Davydov discovered that human psychophysiology has natural analogues, it was necessary to solve the problem of translation of these analogues onto purely human characteristics: character properties, life algorithms, and so on. This problem was successfully solved, and the evidence of this is veracity and exactness of psychological descriptions of human subtypes' life activities, as a result of decryption of 山海經 (Shan Hai Jing), which are created for the Catalog of Human Population.

How are images considered through natural analogues? We think that this will be clearer through an example. Therefore, for example, to determine what psychophysiological qualities are inherent in a woman if one of the images of her individual program is a horse—it is necessary to study natural properties of this image. A horse (Lat. Equus) is an odd-toed ungulate animal, weighs up to 300 kilograms, it is polygamous, herbivorous, it lives on the plains, gallops with a maximum speed of 60 kilometers per hour, etc. This information is then transferred onto a person on six factors, and as a result, we find out the following:

- This woman thinks clearly and expresses her thoughts directly.
- Since horses have a strict hierarchy of females, she is a pronounced leader.
- By nature she has an athletic body, and, as a rule, it is well trained (or looks like it is, even if she does not train on purpose).
- She has a moderate layer of fat.
- She has a tendency to run, and she is not a sprinter, but a stayer.
- She is inclined to have a variety of sexual partners, but at the same time she is a very jealous person.
- Emotionally she is hot-tempered.
- She expresses her emotions openly and unashamedly.
- She is not a good homemaker at all—her home will be about as clean as a horse stall.
- And so on.

However, this is only an example, and we demonstrated only a very small part of what just one image of a person's program can uncover about a human as a carrier of a certain program. Since a human is not an animal that has only one image as the basis of functioning. As already mentioned, the number of images in some human programs might exceed one hundred, and each natural image is first of all considered in terms of its natural analog.

However, it makes sense to mention that the method of analysis of each image of a human program or a manipulation mode through natural analogues is not the only method of obtaining psychological descriptions of a certain person. The main method is work with ancient hieroglyphs, by which these images are recorded in the ancient Chinese monument 山海經 (Shan Hai Jing, Catalog of Mountains and Seas), which turned out to be the Catalog of Human Population. Hieroglyphics that are read using the technology of decryption, which was specially created by Andrey Davydov (the author of scientific discovery of the Catalog of Human Population) gives by far more information about a program and manipulation modes of

Homo sapiens than can be learned from all existing encyclopedias about nature.

However, unfortunately, amateurs are strongly discouraged from attempts to independently decrypt hieroglyphics from 山海經 (Shan Hai Jing), as this skill requires long-term learning and training under supervision of a specialist in the field of Non-traditional Psychoanalysis. Otherwise, attempts to decrypt hieroglyphs, by which human programs and manipulation modes are recorded in this ancient source, can lead to highly undesirable consequences, of which quite severe psychical and somatic illnesses can be regarded as the mildest.

5.9 If developers of the Catalog of Human Population state that *Homo sapiens* is an anthropo-zoomorphic being, bio-program of which is recorded by images of animals, then what is the difference between a human and an animal?

As it was found by Andrey Davydov, the author of scientific discovery of the Catalog of Human Population, a human is not an evolved ape, as Darwinists state; *Homo sapiens* is a biological type that by its natural arrangement is not in any way related to neither monkeys, nor any other kind of animal. Since the basis, on which biological type "human" appears and functions is completely different.

If we consider the structure of psyche of *Homo sapiens*, about which neither Darwin, nor the followers of theory of evolution known nothing, a human being is totally different from any animal. However, of course, in order to state this with confidence, it is necessary to know what human psyche is, how it is arranged and how it works; and, of course, on what basis. In short, everything that lies in the plane of value, which is unknown to anthropologists, psychologists, and so on, called psyche ("soul") of a human[1]. On the one hand, they define psyche as a systematic property of highly organized matter that manifests in a particular reflection of objective reality by a subject[2], i.e., they consider it as a system; while, on the other hand, there was not a single scientific source found that would state that human soul is a system, which is one hundred percent identifiable.

However, in general, psychologists' approach to human soul can be considered correct—it is indeed possible to understand the construct of the soul when considering it as a system. As Sima Qian (II-I centuries BC), the creator of Chinese historical science, stated: "once classified, it can be known." Having identified six factors, of which human psyche consists, Andrey Davydov not just came closer to the soul as a system that is a carrier of information, but also one of the requirements of classification systems was observed—compilation of fundamental sections, which are principally different from each other in their characteristics and criteria.

And, what is the basis? When it comes to a *Homo sapiens*, the basis of his existence, as Andrey Davydov found, are archetypal programs. These programs consist of chimeric formations, while chimeras are expressed in images. The latter is systematized into a scheme, which in a new direction in scientific psychology called Non-traditional Psychoanalysis (which does not using any existing methods in psychology, and is based solely on the decryption of the ancient Chinese monument 山海經 (Shan Hai Jing) that turned out to be the Catalog of Human Population) is known as "individual archetypal pattern."

The difference between a human and an animal is not in that what is commonly believed, for example, possession of the ability to think and articulate speech, work activity, creation and use of tools, alleged "spiritual needs," and so on. The only difference is in that animals are much more primitive in terms of images that are the basis of their life activities. An animal, or any other natural object, has only one image in its biological program. For example: only a monkey, only a horse, only salmon, only a pine, only an ant, only an orchid, only a stone, only gold. However, a human has a natural biological program, which is recorded by an impressive number of natural images, and this program, conditionally speaking, can contain a monkey, and a horse, and salmon, and a pine, and an orchid, and an ant, and stones, and gold, and much more, for example, rivers, lakes and seas, mountains and valleys, grasses and trees, various soils. And, this is far from all.

There are 293 unique human programs described in the ancient source 山海經 (Catalog of Mountains and Seas), and one program can be recorded by over one hundred different images. It is not surprising that a human has a much greater survival rate in nature than any other natural object.

Although only in potential. Without knowledge of his natural program, a person is inferior even to animals in some psychophysiological parameters, for example, strength, agility, speed, visual acuity, etc. Even though it is all there in human potential, without knowledge it will never work. However, this does not make *Homo sapiens* relatives of primates.

[1] "Having accumulated rich experimental material, having developed unique methods of experimental study of psyche, today psychophysics reached an impasse. It lost the very subject of its application. Breaking out of the deadlock is possible in acquisition of scientific interest, namely, in defining its place in solution of the main problem of psychology (anthropology, philosophy): what is psyche." (Zarochentsev, K. D., &

Khudyakov, A.I. (1998). Corr. of RAO L. Verbitskaya & Assoc. Prof. B. Sokolova (Eds.), *Pervyy Rossiyskiy Filosofskiy Kongress: Chelovek – Filosofiya – Gumanizm* [First Russian Philosophical Congress. Human Being – Philosophy – Humanism]: *Vol. 7 Philosophy and Human Problem: The Crisis Of Modern Psychodiagnostics.* St. Petersburg: SPSU Publishing House.)

"To state something fundamental about a modern, civilized human is one of the most difficult and thankless tasks that one can think of since the one who states, is limited by the same conditions and is blinded by the same prejudices, as those about whom he has to express his views." (Jung, C. G. (1994). *Problemy Dushi Nashego Vremeni [Soul Problems of Our Time].* Moscow: Progress.)

[2] "Psyche (from the Greek psychikos - soulful) is a systematic property of highly organized matter, manifested in a subject's particular reflection of objective reality, and includes introjected modeling by him of this reality and regulation on the basis of created model of his behavior and activities. The study of patterns of development and functioning of psyche is the subject of psychology as a science." (Konukhov, N. I. (1996). Slovar-Spravochnik Po Psikhologii [Dictionary-Reference Book of Psychology]. Moscow.)

5.10 Why are there so many animals-chimeras and plants-chimeras in 山海經 (Shan Hai Jing)? How are chimeras related to a human and his psyche?

Existing attempts to give an archetype a form of a particular animal or plant seem quite simplified. *Homo sapiens* already proved by the fact of his survival that he has universal characteristics. This means that a human has the ability to contain characteristics of all bio-forms of nature, comprehend them and successfully survive by using them.

It would be more logical to suppose that archetypes, by which natural human programs are recorded, should be complexed, multi-functional, more universal, and not simplified to a specific archetype of a natural element, whether an animal, a plant, a mineral or some natural phenomenon such as a storm, drought, flood, dawn, etc. And, this supposition was made by Andrey Davydov, the author of scientific discovery of the Catalog of Human Population.

In the 山海經 Shan Hai Jing (Catalog of Mountains and Seas) monument, as Andrey Davydov found, chimeric structures are widely used for recording of human "software." From this, an assumption can be made that creators of humanity got involved in connecting unconnectable things in order to make representatives of this biological type more powerful and mighty than animals, and, as a result of this—amazing creatures that are recorded in 山海經 (Shan Hai Jing) came into being. Natural magnitudes turned into chimeras.

However, with all the fantasticality of described depictions of fauna and flora described in 山海經 (Shan Hai Jing), they are not unique. Almost all cultures preserved zoomorphic beings and unusually existing flora. Painting, sculpture, written records, utensils, and so on—everything has traces of chimerical joining. For example, in 1880 in peatlands of Denmark there was found a vessel similar to a boiler. It dates back to I century BC. It is decorated inside and out with a multitude of amazing depictions of human-like beings and amazing animals. Also, one of the strangest animals, of Celtic origin, is a snake with a ram's head. It is impossible to attribute it neither to gods, nor to people or to anything at all. Chimeras of such kind, as if were taken from a single source and spread over all continents of the world.

As it is known, a chimera is an object that combines parts of different living beings. A well-known example of a chimera is the ancient Egyptian Sphinx.

Sphinx is portrayed differently. For example, in the ancient Egyptian art it is an animal with a body of a lion and a head of a human (less often of a falcon or a ram)—for example, the Sphinx of Hatshepsut (New Kingdom, XV century BC). In the ancient Greek mythology, it is a creature with a woman's head, paws and body of a lion, wings of an eagle and a bull's tail. However, in any case, it is "a composite creature"—that is, a totality of different parts of real animals. The same principle—chimerical—is observed in the case of psychical arrangement of *Homo sapiens*, recorded in the ancient source 山海經 (Catalog of Mountains and Seas), which turned out to be the Catalog of Human Population. However, not only zoomorphic or anthropo-zoomorphic chimeras are found in this ancient monument, but also plants-chimeras, etc.

Principled approach in use of chimeras prompted researcher-sinologist Andrey Davydov to think about the possibility of considering functional psychophysiological diapason of a human through chimerical images. And, as he found out, connections of parts of different animals or plants into one, these fantastic chimeras from 山海經 (Catalog of Mountains and Seas), in essence, are nothing other than formulas; formulas, as the totality of qualities, properties, and functions implanted by nature in representatives of different subtypes of *Homo sapiens*.

These formulas are universal in perception and indestructible in time. Images-chimeras from 山海經 (Shan Hai Jing), the source, which turned out to be the Catalog of Human Population, are "formulas of the soul" for subsequent multi-functional actions of a human in ever-changing environment. And, this is the answer to the so-called "great riddle of the Sphinx."

5.11 Do correctors that influence archetypal complexes of *Homo sapiens*, which are described in 山海經 (Catalog of Mountains and Seas), exist?

Yes, such correctors exist; as it was found by Andrey Davydov (the author of scientific discovery of the Catalog of Human Population), some of the basic correctors that influence archetypal complexes of *Homo sapiens* are geographical zones that represent local ecosystems. They have their own specificity in geomorphological, climatic, biological, and other aspects. Their mutual influence gives us transformation of bio-forms (animals, plants, etc.) of external order.

For example, despite that according to the Russian saying "A pig is a pig even in Africa," externally a Russian "khavron'ya"[1] significantly differs from an African warthog. The same applies to plants. For example, an orchid, which has a huge number of forms that depend, among other things, on the geographical zone of its "residence," but in fundamental characteristics Orchidaceae have the same distinguishing characteristics, which allow to attribute them to the same family.

Something similar can be said about *Homo sapiens*, who belong to a certain ethnic formation. In essence—a representative of any ethnos by his distinguishing characteristics does not go beyond human characteristics; however, in form—there is a diapason of differences, which depends on many correctors, one of which is geographical.

Why is the emphasis on geographical zone? Many ethnic groups consider certain geographical zones their historical homeland and relocation to other areas causes a feeling of nostalgia. Researcher Andrey Davydov found that this occurs in consequence of that location of residence has biological analogues, and, naturally, any deviation of these analogues in other regions leaves a person, who has changed his original place of residence, without the possibility to realize his psychophysiological diapason through unconscious comparing of his archetypal complex with natural biological analogs of his geographical homeland. An individual, a group, an ethnos is left in an informational vacuum, which significantly, for example, reduces survival rate in general and at a new region in particular.

Nationality, place of birth and residence allow adjustment of functional diapason, given by nature to a human, by comparing natural analogues of his birthplace with the individual archetypal complex.

[1] In Russian "khavron'ya" is a word used in common parlance, jokingly instead of the word "a pig."

5.12 When I started learning different images of programs of people from Catalog of Human Population, I noticed that there are natural programs of *Homo sapiens* recorded by a small number of images, and in programs of some people there is a huge number of images. What advantage does a person (human subtype), whose program is recorded by a large number of images, have over a person, whose program contains a small number of images?

Subtypes of type *Homo sapiens* have individual archetypal patterns that vary by volume. Some human programs have, for example, thirty images, while others might contain over one hundred images. However, it is worth noting that the quantity of images, by which human programs are recorded, do not at all make one subtype superior to another subtype or subtypes.

Research that was carried out by Andrey Davydov (the author of the discovery of the Catalog of Human Population) and his colleagues engaged in decryption of the ancient Chinese monument 山海經 (Shan Hai Jing), and their comparative analysis showed that by potential that was implanted in a particular subtype—all subtypes are equal.

Moreover, it is impossible to discuss competition between subtypes. It is the same as in the world of animals: a dog cannot compete with a bird or a fish, and lions are not superior to elephants. They cannot be compared because they are different. In the same way, every human subtype has their own spheres of functionality, their own diapason of abilities.

For this reason, existing ideas in civilization about superiority of one group of people over another is archaism, from the point of view that humanity is a type, in which there are 293 subtypes with different psychophysiological sets of qualities. Superiority of one person (or groups of people) over another can take place only within one human subtype, and not between those people, who belong to different subtypes.

According to information from the ancient Chinese monument 山海經 (Shan Hai Jing), which turned out to be nothing other than the Catalog of

Human Population, people can surpass each other only by one criterion: psychophysiological qualities, psychophysiological abilities that get achieved on the basis of knowledge from the Catalog of Human Population, but are not in any way connected with skin color, nationality, or association with some social group.

Superiority among subtypes within human society can exist only because some use knowledge about themselves and other people from the Catalog of Human Population, and some do not. According to research carried out by representatives of Non-traditional Psychoanalysis—this is not the norm; in principle, competition among subtypes should not exist in human society.

5.13 What are "artificial images?"

Firstly, probably it makes sense to remind of that *Homo sapiens* did not invent anything that is not rooted in natural analogues, and it is unlikely that it is even possible to do. Therefore, contrary to that Carl G. Jung considered archetypes of human unconscious to be derivatives of human imagination (fantasy), research in the field of Non-traditional Psychoanalysis showed that at the basis of functioning of *Homo sapiens* are more ancient archetypes—natural images that are recorded in the ancient monument 山海經 (Shan Hai Jing), which later on transformed into archetypes generated by human fantasies. Human fantasies are that what is called "artificial images."

Prior to the scientific discovery of the Catalog of Human Population by Andrey Davydov in the XX century, without being aware by which natural images human "software" is recorded, civilization produced an incredibly huge massif of artificial images. Artificial images were created to replace natural programs of *Homo sapiens* in the absence of the Catalog of Human Population since, as it was already mentioned, without "software" a human, like the computer, is not able to function.

Every epoch, every culture created and creates its set of images: first there were Apollo or Odin, then Zorro and Batman, and then Shrek, etc. Many modifications of artificial “software” were created without coming close to even approaching the question of that each person has natural “software” given to him by nature at birth.

On the one hand, this is explainable since people had to live on some basis because they differ from animals in that their “software” does not work “automatically” (as it is the case with animals) and must be studied, comprehended, mastered. However, on the other hand, artificial “software” differs from natural just like papier-mâché or plastic apples differ from natural apples.

Civilization involved in creation of artificial images, figuratively speaking, forces people to eat plastic apples instead of real. And, it is not difficult to guess that such "diet" leads each person to imminent death. No one has ever been able to remain alive and well, while drinking gasoline instead of water and replacing all natural foods with synthetic. This "plastic diet" leads to colossally rapid wear of the entire human psychophysiology, premature aging of the organism, and, as a result, death.

However, the result is extended in time, and it is similar to the effect on a person when he takes in daily micro doses of poison. Also, due to these "apples" (artificial images), the potential implanted in a human is realized by about 5% at best.

The essence of the phenomenon called "artificial images" can be expressed differently. For example, using the computer language, it is the worse virus that infects a computer. For some time, this virus simply disrupts normal algorithms of functioning of a system, but as a result, sooner or later these violations make a computer completely unusable. Translated onto human psychophysiology, artificial images as a "virus" cause human organism to work incorrectly, not at full capacity, and make mistakes (fatal ones as well).

Practical results of such failures are a variety of misfortunes that happen to a human: failures, all sorts of problems, illnesses, and old age. Sooner or later, due to these violations, each of which breaks the system called "human psychophysiology" more and more, a human goes out of order, in other words—dies.

When it comes to computers, probably in some cases a good specialist can fix a serious problem or failure. However, in case of a human this occurs without a possibility of recovery. And, in this civilization, it is observable everywhere and can easily be noticed—people rarely live more than 90-100 years. And, this is at best, as human life is usually much shorter. Currently, doctors are not capable of overcoming the problem of death; meanwhile, physiologists state that by nature human body is designed for a youthful and healthy life of a few hundred years.

A human was created as such a powerful being in potential that human psychophysiology can work for some time even when it is infected by a similar "virus." No animal would survive in living conditions that people make for themselves and remain alive in them for some time since at the basis of functioning of *Homo sapiens* nature implanted the power of natural images, and by that provided a huge strength margin to a human being as a living system.

However, a human should understand that if he is still alive, it does not at all mean that he lives correctly. Moreover, it does not mean that he or she will live forever.

5.14 Do people die only because they live on the basis of artificial images instead of images of their natural programs?

Research of the ancient Chinese monument 山海經 (Shan Hai Jing) and other ancient sources related to it clearly points to that death is not a mandatory event for a human; even more so, at an age, in which death in civilization is considered natural order of things.

According to these sources, death is the result of a mistake or mistakes of a human, which caused a fatal damage to him, as a living machine. And, if a person stops making such frequent and fatal mistakes, he can live as long as he wants. In the 山海經 (Catalog of Mountains and Seas) monument such long lifespans are mentioned that it is not worth to tell modern people about them because this will sound too fantastic to them.

Perhaps the only cause of death that cannot be considered a consequence of human mistake is violent death (murder). In all other cases, people themselves are causes of their death, as well as diseases and other misfortunes that happen to them.

And, as it turned out there is only one cause that forces a human to make fatal mistakes—a “virus” called “artificial images.”

Mistakes that led to death of an organism constitute of a wide variety of actions, behavior, human decisions. It is such a wide diapason of causes and effects that it is not possible to discuss this in detail here. However, in a nutshell, people die not according to laws of nature, and certainly not by the principle “God gives, and God takes away,” but first of all, due to lack of information about themselves, about others, about the world, about nature. And, due to their deep ignorance, laziness, stupidity (and insistence on continuing to be stupid), unwillingness to work and study, lack of desire to lead a life of a human instead of that of an animal.

An individual, without being even a human in that sense, in which it is reflected in the ancient sources, imagines himself already equal to the creator, and on this basis thinks that it is no longer necessarily for him to study the world and himself, to learn, to self-perfect—he thinks that he is already great, perfect and mighty.

However, absolutely all qualities listed above are qualities, which a person gets by creating or using artificial images. There is no other reason. People, who live on the basis of artificial images become dull and useless: they do not want to know and understand anything; do not want to do things right, but only in the way that they want; they are not interested in anything, they do not want anything and do not seek anything except pleasure and doing nothing; they are too lazy to take care of themselves, their own health; they violate natural, moral and ethical laws; they engage in self-destruction, as well as in destruction of everything around them.

However, the worst thing for a human is lack of information. Since when a person does not have required information about the external and internal world, he begins to fantasize, imagine. Artificial images are fantasies (including human fantasies about themselves). And, this is the worst thing that can happen to a human since this is the reason various misfortunes happen to him.

Quite often people are so carried away by their fantasies that they simply do not want to see the surrounding reality, even when the information is literally right under their noses. A human thinks in the following way: "Even if I am mistaken, I will continue to think and do what I want, as it is comfortable and pleasant to me." Each person certainly has every right to hold this position. However, from our point of view, this is typical suicide, and in a perverted form; a personally programmed euthanasia of oneself: "Let me die from this, but until the moment of death I will feel good, pleasant."

Humanity can play with their toy called "artificial images" for as long as they want, but people should not forget that an image is not a toy. They must understand clearly that if they think-up monsters, then they become these monsters themselves in reality. With all the consequences, the main and the saddest of which is that they become mutants, and nature is programmed in such a way that any mutations get destroyed.

Even though people to not track and regulate this process due to their ignorance—cells of a human body do not take ignorance into account. Cells do not argue what will be better for a person, as that is not their function. They simply realize (at the physical level) that image, which a person chose for himself. This is exactly how any image works; regardless of whether a person knows about this or not. As a result, a person actually becomes those images, which he chose.

Unfortunately, all of the above is not just another horrifying fairytale of fiction. All of this has scientific evidence: artificial images, instead of natural images, used by a human as etalons of his individuality lead to the state of a mutant, and all mutations in nature get destroyed. Therefore, nature shortens lives of people, who live on the basis of artificial "software" to a minimum, and then at some point turns on the program of self-utilization in them. Nature does not need people, who transform themselves into mutants by their mad fantasies, who do not fit into the environment in a normal, natural way. Mutations—in this case, it is the only reason that people in this civilization die.

5.15 How can awareness help a person avoid damage caused by artificial images?

It cannot since the function called "awareness" does not work in modern *Homo sapiens*. This is stated on the basis of that, according to ancient sources, awareness[1], as a purely human function, is precisely a person's ability to monitor images, based on which he lives and functions.

However, first of all, this function in modern *Homo sapiens* is turned off by artificial "software," based on which they live; secondly, not knowing images of their natural "software" and artificial "software," which civilization "awarded" to them, it is simply impossible to monitor any of the images; thirdly, in order to, as the saying goes, "separate the wheat from the chaff" it is necessary to know the true functions of both kinds of images, and it can be stated with absolute certainty that even the most "advanced" subjects in this civilization simply do not know how to do this. In order to do it, it is necessary to learn how to do it. And, the "first grade" in this "school" is knowledge of images a person's own natural program and his natural manipulation modes.

Without having knowledge neither of your own natural images and what qualities, functionality, abilities and possibilities are behind them, nor of images of artificial "software" (what they are, which invisible programs are behind each of them, how each of them influences you, etc.)—it is impossible a priori to become an individual, who lives consciously, aware.

For a person, who seeks not to be a zombie, but instead to live a life of a sensible human—that is, to have awareness, it is necessary to know that the process of monitoring of images, based on which he lives, is not done by the "all-seeing God," but by his own organism, on the inside. This is how a human is programmed. And, this mechanism cannot be fooled or cheated. A person will have to pay the highest price for his mistakes in regard to his psyche and body, and sometimes the price is very high.

From this it likely follows that the main goal of acquisition of awareness as a quality by a human is to avoid, using the computer language, an early break down and "the blue screen of death." Since this is precisely the price for those, who prefer to live on the basis of their own fantasies. And, first of all, fantasies about themselves, about how they are arranged and how they must live.

And, nature was created so that every person could learn and study his own self through natural images of his individual program, implanted in his unconscious.

Not without reason the Bible states that god created nature for a human. Information from Shan Hai Jing confirms this.

Nature is not only an ideal environment for survival of human species by composition of air, water, etc., food base, pharmacy, a vast reservoir of a variety of natural materials, from which a person can build a house, make clothes, shoes, dishes, tools, and everything else that he needed in his life—nature also provides real living or non-living etalons of what a person must be like, how he must live. On earth it is mountains, valleys, rivers, seas, lakes, trees, grasses, a variety of plants and animals, rocks, minerals, metals, birds, fish, reptiles, insects, flowers, sand, soils, and much, much more. Above it is clouds, fogs, rains, winds, celestial bodies, stars, planets of the Solar System, and so on.

All of these are natural images. And, all that is left to a human is to find out, by which images his personal program is recorded in the ancient source 山海經 (Shan Hai Jing)—the source that by all appearances was intentionally left to humanity for this purpose—and to begin to study them in the environment. Due to existence of these images in the environment, a person can actually see, hear, touch, smell, and feel them. And, through this—understand how these images live, function, look, move, smell, sound, and so on.

Thus, a person receives knowledge of how he himself must live and function, what he should be to follow the co-called "Plan of the Creator," and, most importantly, conform to himself—real, natural.

[1] Modern psychology considers awareness a continuous monitoring of current experiences—that is, a state, in which a subject is focused on experiencing the present moment and tries not to get distracted by events of the past or thoughts about the future. This is the ability of consciousness to introspect its own activities.

5.16 Why do developers of the Catalog of Human Population—Andrey Davydov and Olga Skorbatyuk—consider images the protolanguage?

Because developers of the Catalog of Human Population—Andrey Davydov and Olga Skorbatyuk—get any kind of information about any representative of the biological type *Homo sapiens* precisely from images, which are hieroglyphically recorded in the ancient monument 山海經 (Shan Hai Jing). On this basis it was concluded that natural images are that protolanguage, which linguists have long been looking for.

As it is known, protolanguage is a language that is being reconstructed by linguists—it is the language of the earliest communities of *Homo sapiens sapiens* (modern humans), which had dialects already in Africa 80-160 thousands of years ago. However, if it is also known that a human being, of any race, nationality and language group thinks in images, then—is an image not the protolanguage of the existing population of Homo sapiens?

Words, in any language, are only a form of transmission of mental images. After all, a person, who is deaf since birth does not hear words of any language, but this does not mean that he lacks the process of thinking. Unconscious of a human, his physiology, as well as human cells, do not understand the words of any language.

If a cell would understand the language of words, geneticists would not have the following problem: "How to talk to a cell?" Cells understand each other, but they do not understand a human behind the apparatus for transmission of information to cells. Geneticists need a language that cells understand. However, this problem has not been solved to this day since they already know that human cells communicate with each other by images, but exactly which images need to be transmitted to cells of a specific organism—they do not know.

In conclusion, it is worth noting that the transition from hieroglyphics to phonemographic language left words without images. An image is expressed by a word, recorded by letters of an alphabet, and each person who says it—at the basis has his own, individual image that is different from the image of another person, who uses that same word. The meaning, the essence, the same diapason of understanding with replacement of hieroglyphic writing with phonemographic—by varying degrees destructurizes interpersonal relationships.

The only culture that preserved hieroglyphic writing unchanged is the culture of China. By virtue of this, the ancient Chinese monument 山海經 (Shan Hai Jing), which in spite of everything was also preserved unchanged in this culture, was returned to civilization as the Catalog of Human

Population by its discoverer, researcher-sinologist Andrey Davydov and his colleagues.

5.17 What do images from Shan Hai Jing look like?

The only monument that humanity preserved in its entirety and unchanged is 山海經 (Shan Hai Jing, translated as the Catalog of Mountains and Seas)—the richest collection of natural images that exists in this civilization.

These images seem quite fantastical, and for this reason, already for many centuries, they bring readers and researchers of this ancient monument to an impasse. For example: "...River Shi flows out of there, flows to the northeast and flows into the sea. There is an animal that looks similar to a wolf, but with a red head and eyes like a rat. It grunts like a piglet... There is a bird, similar to a rooster, but with a white head and legs like a mouse, claws like a tiger..."

The entire Shan Hai Jing contains similar material. This ancient monument has only images of similar kind and nothing else. Shan Hai Jing consists of 18 sections, and just 5 chapters related to section 1—Catalog of the Mountains—contain 15,503 hieroglyphs. And, behind most of these hieroglyphs are objects of the surrounding nature. However, all images from Shan Hai Jing are natural images because they are images, on the basis of which psychophysiology of any representative of human population was created and functions. These are not just images—they are images of natural human programs and manipulation modes. And, from these images, it is possible to find out anything you want to know about any person.

PART 2

300 SELECTED RESEARCH TOPICS OF THE SPECIAL SCIENTIFIC INFO-ANALYTICAL LABORATORY—CATALOG OF HUMAN SOULS (1974-2014)

Sketch © 2000 Andrey Davydov

UNIT 1

SCIENTIFIC PSYCHOLOGY AND OTHER SCIENCES

The Catalog of Human Population was found. The answer to the question "What is the "soul" (psyche) of a human and what is its structure?" was found since the Catalog of Human Population was discovered. The basis for cataloging biological type *Homo sapiens* is the discovery of existence of subtypes (subspecies) within this type (species). At the moment of birth, a person is not "a tabula rasa" ("a blank slate")—he is born with a subtype program and three manipulation modes of correction of his psychophysiology and behavior. Manipulation modes can also be used to manage a person from the outside. Programs and manipulation modes of each subtype are endemic.

SECTION 1

SCIENTIFIC PSYCHOLOGY.
NON-TRADITIONAL PSYCHOANALYSIS.

Systematic research in the field of human psychophysiology, which united psychology, sinology, linguistics, phenology, biology, zoology, botany, mineralogy, geomorphology, physical geography, hydrology and other sciences. The Catalog of Human Population is a description of human as a type by subtype structures. According to the definition that was developed and introduced to non-traditional psychology by the author of the discovery of the Catalog of Human Population—Andrey Davydov: "The Catalog of Human Population is a description of type human by subtype structures. Subtype structure (synonyms: "psyche", "soul") is a combination of individual archetypes, recorded at the genetic level (principle). Expressions and interaction of subtype structures in manipulation modes and phonological algorithms are described with adjustments for

gender, age and cultural differences. Information is recorded on six factors."

1.1.1. Psyche is an archetypal matrix.

1.1.2. Complete and detailed information about nature and structure of human psyche was found by researcher-sinologist Andrey Davydov in one of the most ancient monuments, which some researchers date XXIII century BC.

1.1.3. Catalog of Human Population. *Homo sapiens* is a biological type (species), divided into 293 subtypes (subspecies).

1.1.4. A human is a programmed bio-form, an anthropo-zoomorphic being, whose individual (subtype) archetypal pattern depends on the phenological cyclicity of a geographic area.

1.1.5. Three human manipulation modes (suppression, balance, stimulation) are correctors of his psychophysiological state and behavior.

1.1.6. Classification by six factors, which are principally different from each other, allows concretizing the functional diapason of an individual by sections.

1.1.7. Image-hologram carries information compendiously, multi-dimensionally in space and beyond time.

1.1.8. Images as segments-blocks that make up individual archetypal patterns of subtypes structures (psyche) of *Homo sapiens*.

1.1.9. Influence of natural and artificial images on human psychophysiology.

1.1.10. Artificial images as the root of any problem of an individual, as well as society as a whole.

1.1.11. An image is the basis of management of a system called "a human."

1.1.12. Non-traditional psychoanalysis is a new direction in scientific psychology.

1.1.13. From Carl Gustav Jung's archetypes of the collective unconscious to individual archetypal patterns.

1.1.14. Can archetypal images contain chimeras?

1.1.15. Archetype semantics: how it corresponds to the concept of "an image." How archetypal are images?

1.1.16. Self-perfection as a conscious achievement of wholeness of an individual by the means of certain qualitative transformation of individual human subtype program with transition to a concretely new level, criterially recorded in the Catalog of Human Population.

1.1.17. Historical background of the discovery of the Catalog of Human Population.

1.1.18. General information about the author's method and technology of decryption of Shan Hai Jing (and other ancient monuments) and creation of the Catalog of Human Population.

1.1.19. Practical application of the Catalog of Human Population.

SECTION 2

SINOLOGY.
THE ANCIENT CHINESE MONUMENT SHAN HAI JING TURNED OUT TO BE THE CATALOG OF HUMAN POPULATION.

The main resource of knowledge about nature and structure of human psyche turned out to be the ancient Chinese monument Shan Hai Jing; the ancient sources I Ching, Tao Te Ching and others turned out to be annotations for use of the Catalog of Mountains and Seas as the Catalog of Human Population.

1.2.1. The ancient Chinese monument Shan Hai Jing—myths or the structure of psyche?

1.2.2. Catalog of Mountains and Seas is the map of human psychophysiological structure.

1.2.3. Ancient Chinese sources Shan Hai Jing, Tao Te Ching, I Ching and others turned out to be a series of texts on the same topic.

1.2.4. Lao Tzu's treatise Tao Te Ching turned out to be an annotation for the use of Shan Hai Jing as the Catalog of Human Population, I Ching and other ancient sources.

1.2.5. "Dao is not empty": incorrect translations and interpretations of Tao Te Ching make this treatise as a practical guide into an incomprehensible and useless text.

1.2.6. I Ching is not a book, but only the table of contents to a text, which was not found. I Ching as the system of files of an ancient computer.

SECTION 3

SOCIOLOGY AND ANTHROPOLOGY. *HOMO SAPIENS* ARE BIO-ROBOTS. SOCIETY AS A COMMUNITY OF MANIPULATORS AND THEIR SUBJECTS.

A human does not exist in this civilization. This conclusion is based on information from a source, which turned out to be the Catalog of Human Population. The Catalog was created as project documentation, on the basis of which *Homo erectus* was created. A modern person completely groundlessly took the title *Homo sapiens*. *Homo erectus* is a form, which might eventually develop into *Homo sapiens*, or not develop. For millennia an individual of this civilization is in marginal discomfort due to inability to uncover and realize his psychical structure, which is the root of any problem from the category of "human-society." With the Catalog of Human Population any problem from the category of "human-society" is easily solvable. Civilization can think and feel anything it wants about the Catalog of Human Population—this does not prevent it from existing absolutely autonomously.

1.3.1. *Homo "sapiens"* is a bio-robot.

1.3.2. Why with principally identical human psychophysiology qualities of properties are different? 293 subtypes of biological type "human."

1.3.3. Separation based on natural subtype programs gives an opportunity to discard separation by race and to perceive race only as a minor corrector to a program.

1.3.4. The difference between any representative of modern civilization and a human is the same as between a seed and a tree. From a primitive bio-robot to a human.

1.3.5. Humans did not evolve from apes. Attempts to study the subject of anthropology without the Catalog of Human Population are nonsense.

1.3.6. Society as a community of manipulators and their subjects.

1.3.7. Catalog of Human Population: a new scientific method of gaining knowledge about the initial element of the social structure—a human.

1.3.8. The discovery of natural manipulation modes of *Homo sapiens*: the problem of how to make any person act, think, live in a certain way is solved.

1.3.9. Practical uses of three individual natural manipulation modes of a person (suppression, balance, and stimulation modes).

1.3.10. Ethics of manipulation. Today, the use of natural human manipulation modes from the Catalog of Human Population is the only alternative to violence.

1.3.11. Discrimination of any kind is unnatural and anti-natural. Catalog of Human Population is the source of knowledge without any kind of discrimination.

1.3.12. Any problem from the category "human-society" is easily solvable with the Catalog of Human Population.

SECTION 4

INTERDISCIPLINARY STUDIES

The Catalog of Human Population can be used in many fields of scientific knowledge; for example, in genetics. It has been established that cells exchange audiovisual "packets" of information (images), but the language for productive "conversation" with a cell has not been found. In view of this, probably, geneticists would be interested in the topic of "Intercellular imagery information within images described in the Catalog of Human Population." According to the principle of symbiosis, we could suggest topics for joint research in many other scientific fields.

Lectures are created by special request from scientific establishments.

UNIT 2

PRACTICAL PSYCHOLOGY. THE SOUL EXISTS, AND IT IS KNOWABLE.

"The Golden Key," which can open human souls was found. This key is knowledge of natural "software," with which a person was born. A human, like a computer, cannot function without "software." "Software" of a person consists of an individual program and three manipulation modes. No one is able to go beyond their "software." Therefore, it is possible to obtain any kind of information about any person from the Catalog of Human Population, including information about how to manipulate (control) him with a guaranteed result. Each person has his own "software," and it is implanted deep in the unconscious, which is why a person is not able to uncover his own (let alone other people's) "software" without a "textbook."

SECTION 1

SELF-KNOWLEDGE AND SELF-PERFECTION. YOUR SOUL IS NOW A PUBLIC PLACE. LEARN ABOUT YOURSELF BEFORE OTHERS LEARN ABOUT YOU.

Achievement of the state of being your own master and an autonomous individual, wholeness and transformations—are now possible.

2.1.1. Scientists decrypted a very ancient source, which contains 293 individual models of psyche (soul) of *Homo sapiens*.

2.1.2. A human has an individual natural program from birth. Knowledge of the human program is an opportunity "to look behind masks" and find out everything about him/her/yourself.

2.1.3. Every person is manageable. Any person can be manipulated by the use of strictly defined programs.

2.1.4. In the beginning was not a Word—in the beginning was an Image. An image is the language, by which all information about a human is recorded in his unconscious.

2.1.5. The secret of ancient priests. Images are the key to managing any person.

2.1.6. Natural and artificial "software" of a human and its influence on psyche and body.

2.1.7. "Papier Mâché apples": attempts of humanity to "self-program" ended tragically—not one human exists in this civilization.

2.1.8. The great heritage of ancestors. Historical background of the Catalog of Mountains and Seas as one of the most ancient monuments (which some researchers date XXIII century BC).

2.1.9. Historical background of the discovery of the Catalog of Human Population and the story of its researchers.

SUBSECTION 1

HUMAN MANIPULATION MODES: EITHER YOU ARE A MANIPULATOR, OR YOU ARE BEING MANIPULATED.

Management of you is located, figuratively speaking, under your skin. To be more precise: it is much deeper—in your psyche. Any person can be manipulated (controlled) with a 100% guaranteed result. Human Manipulation Modes is a mechanism implanted by nature, by the use of which any person becomes manageable.

2.1.1.1. Manipulation modes are an inseparable part of structure of human psyche. They are given from birth at the level of reflexes, and, therefore, no one can resist their application.

2.1.1.2. Every person has three individual manipulation modes from birth: suppression, balance, and stimulation.

2.1.1.3. Suppression mode: a person onto whom his personal suppression mode is transmitted begins to fully submit to the manipulator at the unconscious level.

2.1.1.4. Balance mode: a person onto whom his personal balance mode is transmitted, unconsciously begins to trust and experience friendly feelings towards the manipulator.

2.1.1.5. Stimulation mode: a person onto whom his personal stimulation mode is transmitted will perform any actions that the manipulator needs.

2.1.1.6. When one's personal manipulation modes are applied, he forms a complete willingness to do even that for the manipulator what he would never have done voluntarily.

2.1.1.7. With application of manipulation modes, a manipulator is not only guaranteed to get what he needs, but is also insured from discontent, negative reactions, resistance of the person he is manipulating.

2.1.1.8. Any individual manipulation mode is someone's individual program. Why it is necessary to know dates of your manipulation modes.

2.1.1.9. Knowledge of your own natural program and your own manipulation modes allows you to become inaccessible to manipulations from outside.

SUBSECTION 2
WHY THE CATALOG OF HUMAN POPULATION IS NECESSARY

2.1.2.1. Benefits, which a person gets from knowledge of his own/someone else's personal natural program, his/others' manipulation modes and those images, by which all this information is recorded in the unconscious.

2.1.2.2. Why the Catalog of Human Population is necessary.

SUBSECTION 3
HUMAN INTELLECT

Despite that human intellect was put on a pedestal, and, moreover, some psychologists consider it to be psyche (soul)—*Homo* sapiens are not "sapiens." And, it is not simply that (according to specialists) they use only a few percent of their

intellect—actually, it is completely turned off. It is turned off by artificial "software." People of this civilization do not think—they live solely on reflexes.

2.1.3.1. An artificial mechanism instead of intellect works in modern humans—that is, standards given by parents and society in the process of upbringing, which completely blocks the work of human intellect.

2.1.3.2. *Homo "reflexus."* When a person does not know his natural program, his intellect does not work and he lives on reflexes, like an animal.

2.1.3.4. That what you consider your "inner voice" is an "installed" virus.

2.1.3.5. What does it mean "to think": when you do not live on reflexes, then it is necessary to think, but people of this civilization do not know what this process is.

2.1.3.6. Finding out his own natural program is one of the areas, where intellect of a modern human failed.

SERIES OF LECTURES. A virtual house, where hearts break.

2.1.3.7. Human "virtual room."

2.1.3.8. The soul as the carrier of natural "software" has nothing in common with that what a person considers his "inner world."

2.1.3.9. Turning to your inner "virtual world" with absolute certainty that it is your soul is the shortest road to madness.

2.1.3.10. The place that every person considers the only "private zone," which no one except him can access, became a loophole for "puppet-masters" a long time ago.

2.1.3.11. Comparative analysis of inner "virtual constructs" and images of natural human "software": a natural image has enormous potential, and in comparison with it an artificially created virtual image is like a nail compared with a computer.

2.1.3.12. Inner "virtual room" is given by nature, so that every person could make himself into a human through the study of his natural images.

2.1.3.13. Why else a *Homo sapiens* needs "a virtual room": renewal of intellect, creation of philosophy of your subtype, etc.

SUBSECTION 4
SELF-PERFECTION

Achievement of the state of being your own master and an autonomous individual, wholeness and transformations—are now possible.

2.1.4.1. Achievement of the state of being an autonomous individual is now possible. What is an autonomous individual and what is the purpose of becoming such an individual?

2.1.4.2. Self-perfection as achievement of wholeness with total transformation is not possible for a person without knowledge of his natural program.

2.1.4.3. A description of any subtype structure from the Catalog of Human Population includes complete information on how an individual can use his abilities to the maximum.

2.1.4.4. "Upgrade": real metamorphoses of an individual are possible only with reconsideration and revision of the philosophy, based on which he lives.

2.1.4.5. Get rid of "viruses": how to "modernize (renew)" yourself independently and with a guaranteed result.

2.1.4.6. "Upgrade": the intellectual factor.

2.1.4.7. "Upgrade": the physical factor.

2.1.4.8. "Upgrade": the nutritional factor.

2.1.4.9. "Upgrade": the emotional factor.

2.1.4.10. "Upgrade": the sexual factor.

2.1.4.11. "Upgrade": the environmental factor.

SECTION 2

A MAN AND A WOMAN

A man, a woman, sexual relations are not at all that what you think.

2.2.1. There are no men or women in this civilization. There are only males and females with psychology of little boys and girls.

2.2.2. You do not have sex. How to make sure that you have sex and that your partner does not treat and use you only as a sex toy.

2.2.3. What can replace men's sexual standards: *"Stick it in, pull it out and run away"*, *"Drain the sperm into a woman, as into a toilet bowl,"* etc.

2.2.4. What can replace women's sexual standards: *"A sexual partner is a living dildo, and in my fantasies I'm having sex with another man/other men,"* exchange of sex for material and other benefits, etc.

2.2.5. What can replace men's and women's sexual standard called *"You are obligated to arouse me."*

2.2.6. In reality, erogenous zones (yours personally and your partner's) are not where you think: their placement is individual and depends on the psychical structure of a person, about which you know nothing.

2.2.7. Application of artificially created images means absence of sex; the use of natural images, by which programs of sexual partners are recorded means presence of sex.

2.2.8. Without the Catalog of Human Population it is impossible to have natural sexual processes, purposeful use of sex, natural orgasmic sensations with various power and duration.

2.2.9. Sex not as satisfaction of other people's sexual fantasies, but as satisfaction of your natural sexual desires.

2.2.10. "Oral, anal and normal": the truth about sexual taboos.

2.2.11. "Wild sex": how to have sex out of images of animals, plants and other images from the Catalog of Human Population, and at the same time get a lot of pleasure.

2.2.12. It is normal to have multiple sexual partners. Origins of monogamy and polygamy are in natural human "software."

2.2.13. For those who want to create a pair for life. Sexual relations, which both partners never become bored of are possible.

2.2.14. Who needs the institution of marriage and family, what for, and is it necessary for you personally?

SUBSECTION 1
FOR MEN

What no one told you about women.

2.2.1.1. Some good news for men: your appearance, age, financial status, and even the size of your penis will not make any difference to a woman if you act out her manipulation modes.

2.2.1.2. Using the tool "natural human manipulation modes" any man is able to not only incline any woman to have sex with him, but can also make her his lover without having to "butter her up" by money, gifts, etc.

2.2.1.3. Who is a man for a woman? He is a resource.

2.2.1.4. Women's thirst for power is much stronger and more stable than men's.

2.2.1.5. A woman is a cruel exploiter in sheep's clothing, she sends a man "to slaughter."

2.2.1.6. Women always lie. A woman who tells the truth is sick.

2.2.1.7. Women's secrets are not such an innocent thing, as it is commonly thought—it is a powerful weapon against men.

2.2.1.8. One of women's secrets is professional use of emotions as an emotional "dose" or emotional "syringing."

2.2.1.9. Women's tactic: "I'm stressed because of you, give me some money!"

2.2.1.10. Men—boomerangs. How and what for women "launch men into orbit."

2.2.1.11. A woman is the most cruel family member. Women's emotional experiments are psychological terrorism.

2.2.1.12. All women are cruel avengers when it comes to men; they take revenge even on those men, who have never done anything but good for them.

2.2.1.13. Women lie that only men need sex.

2.2.1.14. "Goddesses of fate" (women) fully control men's lives: birth, death and the time interval between these events. They model, moderate, manipulate.

2.2.1.15. Women have a firm opinion that "all men are pigs." Where did "real men" go?

2.2.1.16. Revelation of the secret of why women sometimes withhold sex from men.

2.2.1.17. Now any man can not only have sex with any woman without any expenses, but also, with the help of skillful manipulation, can incline a woman to share her financial assets with him.

SUBSECTION 2
FOR WOMEN

How men turned women into nothings and got total power over them by creation of artificial images. Legalization of women's secrets.

2.2.2.1. With application of a man's natural personal manipulation modes—parameters of a woman-manipulator absolutely do not matter: how old she is, whether she is attractive or not, etc.—she will be love anyway, and men will do whatever she needs them to do and will be by her side as long as she wants.

SERIES OF LECTURES. A new toolkit for women to manage men.

2.2.2.2. Mother's/grandmother's life recipes are the root cause of all women's problems: from pimples to loneliness or cancer.

2.2.2.3. Women's clichés are keys, which like thieves' "lock-picks," are intended to break into the psyche of another person.

2.2.2.4. Cliché Actress: Those women who do not know about this, play the role of "sexy," look silly and vulgar, and are unable to make a man or another woman fully submit.

2.2.2.5. Cliché Clown: Strong impact on psyche of another person with a sweet smile on her face.

2.2.2.6. Cliché Shaman: Use of this cliché allows a woman to master her skill of seduction to such an extent that she could cause an erection even in a deadman.

2.2.2.7. An alternative to seduction of people of either gender by the use of traditional methods: individual natural human manipulation modes.

2.2.2.8. Women do not know that emotions can be a fountain of life, as well as a fountain of death, their death.

SERIES OF LECTURES. The method, by which men turned women into nothings.

2.2.2.9. How by the use of a secret method of influence through images, men managed to get complete power over women.

2.2.2.10. World history of turning women into mats.

2.2.2.11. "All women are the same." Artificial images—a tool for mass unification of women.

2.2.2.12. Who, how and what for turned a woman into a free add-on to her vagina.

2.2.2.13. For "princesses": how favorite since childhood heroines of fairytales break women's lives, turning them into trash.

2.2.2.14. *"I already know everything"*: how the image of the biblical Eve turns women into mentally retarded and stupid.

2.2.2.15. *"I eat like a bird"* is the best recipe to gain excess weight, and the dream *"Be near me always"* is the way to an intensive care unit. Why it is detrimental for women to know nothing about the influence of images on human psychophysiology.

2.2.2.16. If by nature a woman is not a clone of a Barbie-doll, then who? And, how to become a woman?

SERIES OF LECTURES. A woman is a log with legs spread wide. How goddesses were turned into logs.

2.2.2.17. Is it true that while a woman is in demand as a female, she is interesting to nature as a biological species?

2.2.2.18. "All women are stupid": the reason why a woman's intellect dies and how it can be reanimated.

2.2.2.19. The deeper you get into your fantasies, the closer you get to a mental disorder and a shovel to dig a hole with and bury yourself.

2.2.2.20. Women and emotions: "dead and beautiful." Why women forbid themselves to follow their emotions as natural needs.

2.2.2.21. Is it necessary for a woman to take "molds" of a man in order to amuse herself with fantasies that she knows, understands and controls him?

2.2.2.22. A girl and a doll. A simple and effective recipe for making any woman more primitive than a toaster and easily managing her.

2.2.2.23. An image is a magic wand that shoots from both ends simultaneously. How women unintentionally kill themselves.

2.2.2.24. How and why a woman turns herself into scum.

2.2.2.25. Men try to find themselves, while women look for something for themselves.

2.2.2.26. A woman's thirst for power over a man is the way to become garbage.

2.2.2.27. All women (by their psychology) are card tricksters.

SERIES OF LECTURES. "Head in underpants."

2.2.2.28. Why women's heads are "in their underpants" for most of their lives.

2.2.2.29. Women's promiscuity (as well as men's) is neither a disease nor a vice. How to be a slut, and with pleasure.

SERIES OF LECTURES. Business women's way.

2.2.2.30. Business women's way is a short time period between pads and condoms.

2.2.2.31. How to arouse a boss or a colleague of either gender (during business hours as well) without looking like a prostitute.

SERIES OF LECTURES. Feminism.

2.2.2.32. Feminists do not really need that what they are fighting for because that is also forced upon them by men.

2.2.2.33. Women do not need equality with men—women need power over men. With the Catalog of Human Population it is now possible to get such power.

SUBSECTION 3
GAYS, LESBIANS.

Undeservedly accused and unjustly persecuted.

2.2.3.1. It is natural and normal to be gay or lesbian. Although not for everyone.

2.2.3.2. For gays and lesbians: ways to overcome negative attitudes towards yourself, as well as internal psychological problems.

2.2.3.3. A method for finding sexual partners without having to deal with those people, who are not inclined to have homosexual relations or have strong negative views on homosexuality.

2.2.3.4. Fundamentals for creation of an ideology for leaders of organizations for gays and lesbians.

SUBSECTION 4
"SEXUAL PERVERSIONS"

It turned out that most "sexual perversions" are norms.

2.2.4.1. Which specifically so-called "sexual perversions" are norms, for whom and why this is so.

SUBSECTION 5
DATING, RELATIONSHIPS, FAMILY, SEARCH FOR A PARTNER.

Loneliness and failures in love are problems only for those, who do not use the Catalog of Human Population.

2.2.5.1. Love is not something you need to dream about your whole life—it is something you can have right now and always.

2.2.5.2. The root cause of loneliness and problems in love is only that an individual is poorly informed on how to quickly and surely find a suitable partner (partners), and how to become loved forever.

SERIES OF LECTURES. A powerful weapon "in a fight for love"—human manipulation modes: suppression, balance stimulation.

2.2.5.3. Manipulation modes: a simple and easy way to seduce absolutely any man or woman.

2.2.5.4. In order to make someone experience true love towards you, (all the way to adoration, deification, worship) it is necessary to know that person's Suppression Mode.

2.2.5.5. Balance Mode is the shortest way to make a person trust you, experience friendly feelings towards you, and feel comfortable around you.

2.2.5.6. Stimulation Mode motivates another person to perform any actions you need him to perform.

2.2.5.7. How to use manipulation modes to rid yourself of communication with a lover (spouse, relative, etc.) you are bored with or who is angry, and do it easily, quickly, and without detriment to yourself.

2.2.5.8. How to use human manipulation modes to make a partner stop cheating on you even if he/she is polygamous by nature.

2.2.5.9. How to get rid of a competitor by using knowledge of manipulation modes of your partner.

2.2.5.10. How to permanently stop being the object of manipulation in relationships through knowledge of your own manipulation modes.

SERIES OF LECTURES. Ideal partnership is possible.

2.2.5.11. How to find a perfect life partner, with whom you will not have to play any roles, lie, manipulate, etc.

2.2.5.12. How to find a person, who will adore you, worship you like a deity and you will not have to do anything for this.

2.2.5.13. How to find a person whom you will adore, and thus find out what are real passion and true, sincere love.

2.2.5.14. How to find an ideal partner for business, recreation, hobbies, entertainment, creativity, learning, communal living, achievement of some goal, etc.

2.2.5.15. How to find a person, who will quickly (and absolutely for free) relieve you of a bad mood, depression, worries, anxiety, psychological discomfort, and sufferings.

2.2.5.16. How to find a person, who will quickly put you in a state of balance, harmony and comfort without psychologists, psychiatrists and medication.

2.2.5.17. How to find a person, who will almost instantly put you in a state of complete readiness, stimulate you to perform any actions, if you are lacking energy or are too lazy.

2.2.5.18. For businessmen, who make money in the dating industry, etc.: how to take clients "by the lower tie" and not lose them.

SECTION 3

EDUCATION AND UPBRINGING. PARENTS AND CHILDREN. CHILDREN ARE NOT ANTENNA-LIKE CONTINUATIONS OF THEIR PARENTS.

With the discovery of the Catalog, it was found that, unlike it is with plants and animals, children are not antenna-like continuations of their parents. Upbringing of children is a non-existent problem, if you use the Catalog of Human Population.

2.3.1. Children are not antenna-like continuations of their parents.

2.3.2. "*Listen to your mother—kill yourself!*" What and why happens with a child, who follows parental instruction "*Do as I do!*"

2.3.3. Benefits of knowing your child's natural program.

2.3.4. Upbringing as training, which results in a damaged individual, or upbringing as helping the development, which helps a child become a normal adult.

2.3.5. Ahnenerbe—your killer is under your skin. ***(Available in Part 3 of this book.)***

2.3.6. The method, by the use of which mothers get full, uncontrolled access to psyche of their children.

2.3.7. How not to spoil relationships with children in the process of upbringing and not to be left alone without help and support in old age.

2.3.8. Possibilities of application of the Catalog of Human Population in upbringing.

2.3.8. Possibilities of application of the Catalog of Human Population in the education system.

SUBSECTION 1
PROBLEMS OF ADOLESCENTS

Sigmund Freud was right: all problems are rooted in childhood. And, not only psychological problems, but purely physiological ones as well. However, he incorrectly identified origins and the mechanism of this phenomenon.

2.3.1.1. "Problems of adolescents" (loss of control over their behavior by parents and teachers, poor performance at school and lack of desire to study, weight issues, drugs, random sexual contacts and early pregnancy, suicide, etc.) do not exist with the Catalog of Human Population.

2.3.1.2. If a teenager is not getting authentic information on questions "Who am I?", "What am I?", "What am I living for?", "How should I live?"—then, the "Image of Self" that he creates will be a program of nonhumans.

2.3.1.3. How to organize rehabilitation of adolescents, who are serving time for committed crimes and how to help them avoid becoming hardened criminals.

2.3.1.4. How a teenager can easily, quickly and unmistakably choose his/her future profession without contacting "experts" (and with or without help of parents).

2.3.1.5. It is not the beginning of menstruation that causes a psychical trauma for young women, as it is commonly believed—actually, it is something else.

2.3.1.6. At some point, every teenager forms a negative attitude (all the way to hatred) towards their parents, but with current methods of upbringing this is impossible to overcome.

SUBSECTION 2
FOR MOTHERS

2.3.2.1. Why at some point all mothers start hating their children, and how to avoid this.

2.3.2.2. With what to replace "rotten childhood happiness."

2.3.2.3. How and why mothers' untransformed unconscious desire to be needed by someone kills women from the inside.

SECTION 4

BUSINESS, PRODUCTION, MANAGEMENT.

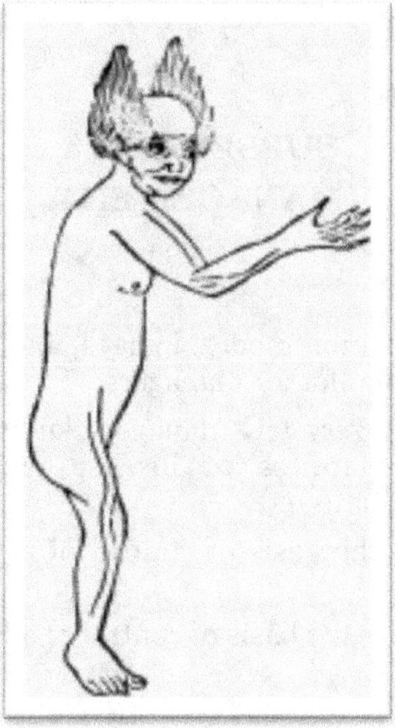

It is possible to turn a human into a 100% controllable producer and consumer.

2.4.1. How it became possible to catalog humanity by subtype characteristics thanks to the technology created by researcher-sinologist Andrey Davydov, and what is the practical use of the Catalog of Human Population in business.

2.4.2. How a businessman can obtain necessary information about people using a minimal (standard form) dataset.

2.4.3. The technology that allows uncovering of any person's individual natural modes of control.

2.4.4. The recipe to outperform any competitors in business on the basis of knowledge from the Catalog of Human Population.

2.4.5. How to effectively sell any goods and services using knowledge of individual (subtype) structures of psyche of clients' (customers').

2.4.6. For global scale "puppet-masters": how to turn a human into a 100% controllable producer and a 100% controllable consumer.

SUBSECTION 1
MANAGEMENT

2.4.1.1. Human manipulation modes, images, and so on—a fundamentally new and a very effective toolkit for managers.

2.4.1.2. The secret that very few managers know: what allows to easily manage any person regardless of his/her race, nationality, gender, educational level, social status, etc.

2.4.1.3. Sensations are the basis of control of a human. Images are the basis of sensations.

2.4.1.4. Which images are the basis of control of a human.

SUBSECTION 2
HUMAN RESOURCE, PERSONNEL AUDIT (HR).

2.4.2.1. An alternative to the method of "trial and error" in staff recruitment is the Catalog of Human Population, which has detailed information about any person as a representative of subtype of *Homo sapiens*.

2.4.2.2. A fundamentally new and a very effective way of working with personnel in a company of any size.

SUBSECTION 3
PRODUCTION

Potentially, there is a huge number of goods, which never yet occurred to anyone to produce, and which are guaranteed to be purchased by consumer.

2.4.3.1. In the sphere of production, 99% of problems occur only due to ineffective actions of a human. How to solve this problem?

2.4.3.2. The discovery of the fact that human species as a type is divided into 293 subtypes (subspecies) requires a fundamentally new approach to production of any goods for a person. What should this approach be?

2.4.3.3. Ideas for production, which have not occurred to anyone, but can make any product that everyone is already tired of attractive and highly demanded: computer games, look-and-feel of mobile devices, mobile applications, clothing, footwear and accessories, interior and household items, cars, printed products, jewelry, adult toys and much more.

2.4.3.4. For producers of printed goods: how to make trillions on certain types of books (posters, calendars, stickers, etc.), which very few people buy today and every person will buy.

SUBSECTION 4
PROFESSION, WORK, CAREER.

Choice of profession or how to win over your boss—are problems, which no longer exist.

2.4.4.1. Should you spend years and money preparing for an occupation that does not fit you, or simply read in the Catalog of Human Population, which professions do fit you?

2.4.4.2. The problem of how to win over your boss no longer exists because the Catalog of Human Population contains information about how to be liked by anyone, and not only that.

SUBSECTION 5
ADVERTISING, IMAGE-MAKING, MASS MEDIA.

There is no need to think-up images. At least because a human is unable to create an artificial image that would have the same power as a natural image. However, there are images that give real power over a human, of which you know nothing.

2.4.5.1. The real reasons behind failures of advertisers, image-makers and mass media in their attempts to influence the audience.

2.4.5.2. Images: real power over a person is possible by having power over his psyche as a whole, not just over his intellect because intellect is not psyche—it is only a part of psyche.

2.4.5.3. Which images have a guaranteed influence on human psychophysiology, regardless of race, nationality, gender, place of residence, profession, social status, etc.

SUBSECTION 6
SMALL/MEDIUM BUSINESS

2.4.6.1. The problem "I work, but I do not have enough money" is solvable. Money is not something that is earned by working. What is money and how to capture it.

SECTION 5

HEALTH, BEAUTY, YOUTHFULNESS.

Health, beauty and youthfulness are achievable with the Catalog of Human Population. (Nutrition, fitness, sports, cosmetics, etc.)

2.5.1. Study your body.

2.5.2. You will not find out how to become young and beautiful from the Internet, magazines or those who call themselves experts, as there is no knowledge about individual programs of *Homo sapiens*. You do not believe this? Wait until you are 40 years old and take a look at yourself in the mirror.

2.5.3. Organics of your soul: your stomach needs organic tomatoes, while your psyche can be "fed" artificial crap?

2.5.4. Are you tired of wasting money on psychologists? In Non-traditional Psychoanalysis, a psychologist knows much more about you than you know about yourself without ever talking to you, or even meeting you.

2.5.5. What your parents, doctors, trainers and nutritionists "forgot" to tell you. What does health really mean and what is the secret of achieving it?

2.5.6. The only alternative to wasting money on doctors, psychologists, trainers, nutritionists, etc. is knowledge of your natural individual program.

2.5.7. Food as medicine: weight of a pill is measured in milligrams, not kilograms, then why do people consume food that they simply like, and take in not milligrams, but kilograms?

2.5.8. With the Catalog of Human Population every person gets an opportunity to choose: to eat like a pig, together with everyone else from a common pan called "a cafe", "a restaurant", etc., or to eat food that nature intended for you personally.

2.5.9. For people with normal weight, who limit themselves in choice of food: how to eat as much as you want and at the same time never have a problem with being overweight.

2.5.10. For people who have weight issues: how to get rid of this problem forever, without spending money on doctors, without pills, without unpleasant procedures and unpalatable diets.

2.5.11. "Legends and myths on nutrition": "Salt and sugar mean white death", "fat is bad," etc.—are not true because, for example, without enough salt a person becomes mentally retarded and might die.

2.5.12. "The great mystery": types of fitness, which are not in your natural program will never make you healthy and beautiful, and sport might make you disabled.

2.5.13. Simple and accessible to everyone: only cosmetic products (for skin, hair, etc.) created using ingredients specified in the natural program of a person—work and do not harm.

2.5.14. Biologically active additives: it is now possible to find out, which ones are right for you without spending money on advisers and unsuccessful purchases.

2.5.15. How to solve problems of alcoholism and drug use without doctors, psychologists and psychiatrists, hospitals and pills.

2.5.16. Psychiatrists cannot heal—they can only turn a patient into "a vegetable" (with subsequent utilization) by the help of psychotropic drugs. What is the alternative?

2.5.17. How to make money in the health-youthfulness-beauty industry using information about clients' individual natural programs from the Catalog of Human Population.

2.5.18. For sports coaches: about how a mere name of a sports team can forever put it in the list of losers, and about many other things.

2.5.19. For restaurant owners: ideas for business, which no one realized yet.

SECTION 6

LEISURE, ENTERTAINMENT.

What do you prefer: a sea of pleasure, unforgettable adventures, incredible impressions and exclusive leisure-time with the Catalog of Human Population, or wasting money on frustration and boredom, like everyone else—the choice is yours.

2.6.1. Should you entertain yourself like everyone else does, with boredom, risks, damage to yourself in order to get at least some impressions and sensations, or should you do it safely and with guaranteed satisfaction?

2.6.2. Leisure-time called "entertainment of *Homo sapiens* subtypes" from the Catalog of Human Population can be very expensive or affordable, but it is always not from the common/for the masses category.

2.6.3. How a businessman can make money in any sector of the entertainment industry on the basis of information about clients from the Catalog of Human Population.

SECTION 7

PROBLEMS OF THE ELDERLY

Instead of a trash can—onto a pedestal. "Subtype dynasty" makes any elderly person demanded, interesting and needed by others again.

2.7.1. How to prevent the loss of a person's desire to live, which turns on the program of self-destruction inside him, and leads to death?

2.7.2. Knowledge of your natural program from the Catalog of Human Population allows an elderly person to feel young again, which beneficially affects his psychical and physical health at any age.

2.7.3. "Subtype Dynasty": what it is and why it makes an elderly person demanded, interesting and needed without any efforts on his part.

2.7.4. For individuals and organizations: how not to help the elderly to die, but how to help them live, and to live with pleasure.

SECTION 8

ARTS, CREATIVITY.

By nature every person is talented and potentially a creator. Any person has a potential ability to make genial creative products in a multitude, but not every person becomes prominent, famous and rich using that what was given to him by nature. There is only one reason for this: his whole life a person tries to find out what his natural talents are and how to realize them in the form of a creative product, and usually there is no time and energy left for realization of his abilities.

2.8.1. How to become not just a creative individual, but an outstanding creative individual, and at the same time make a lot of money on your creations.

SUBSECTION 1
FILM AND THEATER, MUSIC, VISUAL ARTS, LITERATURE AND POETRY.

Secrets of how to make yourself a genial creator.

2.8.1.1. For actors: entry into images using the method from ancient sources is the secret of how to make yourself an outstanding film/theater actor.

2.8.1.2. The secret of how to create a truly unique visual image (painting, graphics, etc.), which are guaranteed to be in demand by buyers.

2.8.1.3. Secrets of creating genial literary works, and in a multitude.

2.8.1.4. Ancient mystical basis for creation of poems.

2.8.1.5. Music as noise or music that powerfully affects the entire human psychophysiology.

SUBSECTION 2
MARTIAL ARTS

How to get qualities and might, and make yourself an unbeatable fighter by using your natural images.

2.8.2.1. How to quickly and easily determine which of type of martial arts and styles fit you personally.

2.8.2.2. How to become an unbeatable fighter using knowledge of images of your natural individual program from the Catalog of Human Population.

2.8.2.3. Martial arts as a method for inner transformation of psychophysiology for those fighters, who use internal styles.

2.8.2.4. For instructors: selection and training of fighters on the basis of information from the Catalog of Human Population regarding who has or does not have which natural abilities.

SUBSECTION 3
DESIGN

Cars, web-design, computer games, interior and garden, clothing and shoes, accessories, hairstyles and nails, tattoos, etc.: all this can make another clone out of a person or make him stand out from the crowd.

2.8.3.1. How and why a traditional, ordinary and seemingly innocent design imperceptibly, but unavoidably kills you.

2.8.3.2. How to stop being a clone and by yourself create unique designs of everything that you use: cars, clothes and accessories, interior, jewelry, and so on based on the Catalog of Human Population.

2.8.3.3. How to become a demanded, well-known and well-paid designer in any field (auto, web, clothing, accessories, interior and garden, etc.) without making special efforts—only by having information about clients' images from the Catalog of Human Population.

2.8.3.4. An elite tattoo is a tattoo with images of your personal unconscious. How clients can get a tattoo that they will never see on anyone else, and which will give them mystical power, while designers will never have to think about what to suggest to a client, so that he orders their services.

2.8.3.5. For producers and consumers of amulets and jewelry: how the former will get huge profits from production of unique amulets and other kinds of jewelry based on the richest resource of images, and the latter—personal products that will not only decorate them, but also will make them healthier, stronger, more powerful, etc.

2.8.3.6. Any professional programmer can gain worldwide fame and popularity of his product by developing computer games and applications on the basis of the Catalog of Human Souls.

SECTION 9

A HUMAN AND SOCIETY.
SOCIETY AS GROUND FOR REALIZATION OF NATURAL QUALITIES AND FUNCTIONS OF AN INDIVIDUAL, AND SOCIAL VALUES AS A SECONDARY CRITERION OF ACHIEVEMENTS.

Firstly, now society is either a prison (and/or an executioner) for an individual or a free supermarket and ground for realization of his natural (program) qualities and functions. Secondly, unnoticeably to the general mass of people, serious changes occurred in society: now every person is either a manipulator, or a subject of manipulation.

2.9.1. How to turn society into a free supermarket for yourself.

2.9.2. Keys to society—"drivers."

2.9.3. "Parallel worlds exist": about people who use the Catalog of Human Population, and those who do not know anything about it.

2.9.4. Society as ground for realization of natural (program) qualities and functions of an individual, and social values as a secondary criterion of achievements.

2.9.5. "You live on a loan": one bio-robot more, one bio-robot less.

2.9.6. The main secret of this civilization is that it is a kindergarten of over 7 billion people, where no one ever grows up.

2.9.7. "Always smile!" and you will get a heart attack. How slogans created on the basis of lack of knowledge of mechanisms of work of human psychophysiology damage and kill.

SECTION 10

CULTURE.
CULTURE IS THE BASIS AND DEGREE OF SURVIVAL OF AN INDIVIDUAL, AS WELL AS A WHOLE NATION.

Culture, in its true meaning, does not exist in this civilization because at the heart of culture must be knowledge about a human and structure of his psyche.

2.10.1. Culture is the basis and degree of survival of an individual, as well as a whole nation. However, there is no culture in this civilization because a person is not given knowledge about fundamentals, basis of his life and functioning.

2.10.2. Myth as consciously perceived reality. As it turned out, myths can contain answers to questions about a human, his soul and functional particularities of his life activity.

SECTION 11

POLITICS, IDEOLOGY.

Modern politics (global and at a state level) is another indicator of a very low level of development of this civilization.

2.11.1. Regulation, control and management of individuals and ethnic groups on the basis of the Catalog of Human Population.

2.11.2. Creation of an ideology for a group of any size is not a problem with the Catalog of Human Population. It is possible to assemble any number of people for any idea that is advantageous to a politician or an ideologist.

2.11.3. Nations that will not use the Catalog of Human Population will be automatically assimilated.

2.11.4. Variants of use of the Catalog of Human Population in politics.

2.11.5. Influence of psychophysiological specifics of a leader on the style of political decision-making.

2.11.6. Elitist political concepts.

2.11.7. Essence of political ideologies and their role in the historical process.

2.11.8. Power as the goal of politics, and what is true power.

SECTION 12

PHILOSOPHY. PHILOSOPHY AS RECIPES OF LIFE VS. PHILOSOPHY AS FANTASIES AND EMPTY VERBIAGE.

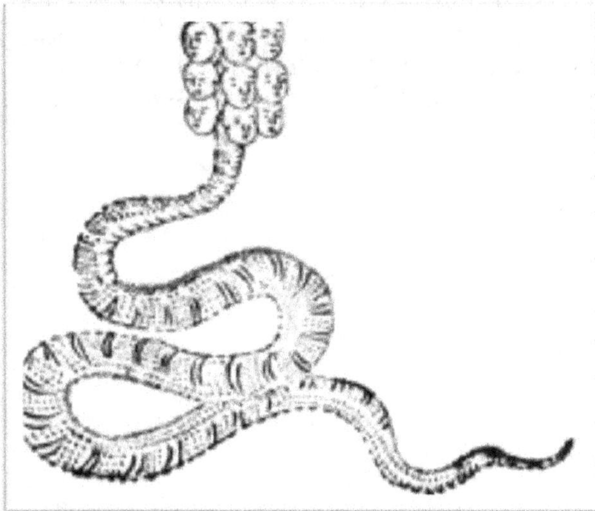

Answers to questions: "Who and what am I?", "What is the purpose of my life?", "How should I live?" and so on, together with detailed recipes of life for each of 293 human subtypes were discovered in the Catalog of the human population.

2.12.1. Philosophy and "philosophy": Catalog of Human Population as a collection of life recipes and an alternative to empty verbiage and uncontrolled imagination of philosophers.

2.12.2. Philosophers now have something to do: true philosophy is philosophy of human subtypes, which currently does not exist in civilization.

2.12.3. Euthanasia instead of happiness, or how philosophy as recipes of life is connected with achievement of state of happiness by a human.

SERIES OF LECTURES. The tragedy of this civilization is that people use words without understanding their true meaning, and moreover, build their philosophy on this basis.

2.12.4. The tragedy of modern civilization is in the absence of monosemantic, authentic definitions of basic (including philosophical) concepts. How this problem is solved through the study of hieroglyphic languages?

2.12.5. What do "good" and "evil" really mean? *(Topics about what are, for example, "health", "power", "money", "desire", "pleasure", "truth", "lie", "fraud", "vileness," etc., with descriptions of mechanisms as recipes of realization of meaning of these concepts—are available by special request.)*

SECTION 13

RELIGION.
THE CATALOG OF HUMAN POPULATION IS THE SCIENTIFIC PROOF OF EXISTENCE OF CREATOR.

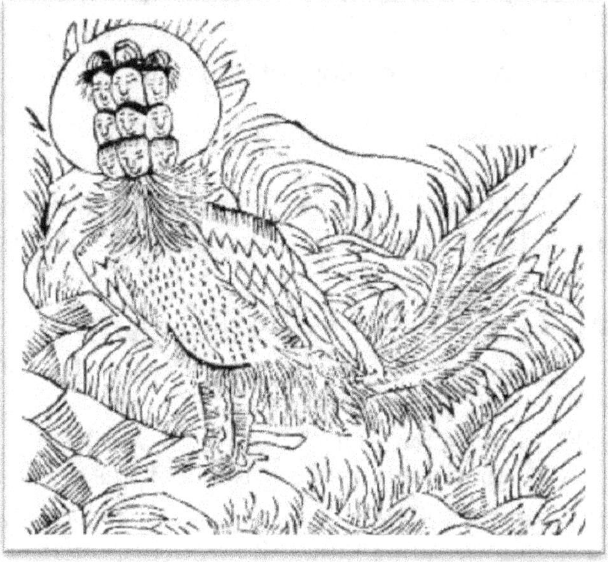

Today, people do not believe in "God," regardless of what they declare. People's real lives, their behavior, their true goals and objectives have nothing to do with the creator's plan. However, the creator (or creators) exists and there is evidence in the form of the Catalog of the human population (also known as the Catalog of Human Souls). The fact of discovery of this Catalog among very ancient (dated XXIII century BC by some researchers) monuments can become the basis of religions, even though it was created at a time when religions did not exist, before the "Bible", "Torah," etc.

SERIES OF LECTURES. The crisis of religion and how to resolve it.

2.13.1. Today, no one really believes in "God". "God" turned into "Santa Claus," in whom people would like to believe, but there is no basis for this.

2.13.2. In order to return spirituality, as the foundation of human life, to this civilization, there needs to be not belief that humanity was created, but knowledge that this is really true.

2.13.3. The basis for a new ideology of religions (Christianity, Judaism, Islam, Buddhism, etc.) is the Catalog of Human Population, as a source, which is the proof of existence of the creator (or creators).

2.13.4. Justifications for new ideology of religions.

SERIES OF LECTURES. No one has yet canceled "the end of the world," as annihilation of this whole civilization.

2.13.5. At the present moment, the future of humanity is its complete annihilation due to unconformity with the creator's (creators') plan, and here is evidence of this in ancient sources.

2.13.6. Spirituality and "spirituality." Return of spirituality to humanity is the only condition for continuation of existence of this civilization.

2.13.7. Responsibility of religious institutions to the civilization, and the possible role of religion in saving humanity from annihilation.

2.13.8. For leaders of religious organizations: how to convince people that "God" exists; the ideology, which is able to greatly multiply the congregation of any of religious confessions.

2.13.9. "Paradise" is here, "God" is near. How a person can obtain an individual "way of salvation," without turning to religious institutions.

SERIES OF LECTURES. What you will not hear at a church. For all those, for whom words 'soul', 'god', 'spirituality' are not empty sounds.

2.13.10. Soul exists. And, the myth of incomprehensibility of a human and his soul was dispelled.

2.13.11. True faith is really knowledge.

2.13.12. *"Memento Mori"*: you were cheated—soul of a human is not second hand.

2.13.13. In the beginning was not a Word, but an Image.

2.13.14. The real meaning of words "in image and likeness."

2.13.15. "Second Coming" will not occur because everything was already said and demonstrated to humanity; Jesus Christ is not a clown.

2.13.16. A believer gets demons instead of the "Holy Spirit" as a result of him being poorly informed about what Spirit actually is.

2.13.17. Actually, the way to Spirit is to study and to learn.

2.13.18. Barbie doll, Shrek and Superman—Sodom and Gomorrah of today.

2.13.19. What really constitutes "a sin" of a human before our creator (or creators).

2.13.20. It is possible to not only have true faith in "God," but also to start getting real support from so-called "Higher Powers."

2.13.21. Does a human need religion? For those, who want to believe in "supernatural powers," and want to get at least some use and benefits out of their faith.

2.13.22. The search for recipes of how to become a being, who is called "God" in this civilization, and the results of this search.

SERIES OF LECTURES. Dead souls: do not look for demons on the outside—they are inside you.

2.13.23. "Demons", "devils", "evil spirits" are those, who modern people actually create themselves, within their own selves.

2.13.24. "Gui" (a person with a dead soul) or, using the language of science, a mutant—is a product of this civilization.

2.13.25. "Demon's head": construction of perversions, which in this civilization are considered genius for some reason.

2.13.26. "Sleep that is death": a person who lives like a computer that is in "sleep mode" is akin to a dead person—inside him is ringing emptiness because his soul has died.

2.13.27. Not "sleep of the mind produces monsters," but sleep of the soul.

2.13.28. "To be yourself" means to be a corpse.

2.13.29. "An elephant and a pug": not "Human and nature," but "Nature and human."

2.13.30. Genial perverts.

2.13.31. *"...Ye shall know them by their fruits"*: indications that a person has a dead soul.

2.13.32. "Affairs of the dead" and affairs of the living. Resurrection of soul is possible.

SECTION 14

ESOTERICISM, SPIRITUALITY.

Secret knowledge does not exist—there are only children, who do not know how to read. Many try to engage in spirituality, but it is meaningless because spirituality itself engages in a human.

2.14.1. Secret (esoteric) knowledge does not really exist—there are only children, who do not know how to read.

2.14.2. Magic and "magic." Magic is a method that gives results, which can be seen, heard, felt; and, "magic" is simply entertainment for the profane.

2.14.3. The main tool of magic is an Image.

2.14.4. An image is the basis of all things. The mystical power of nature.

2.14.5. Magic on the basis of images from the ancient monument Shan Hai Jing is a tool for influencing your own and other people's psychophysiology.

2.14.6. Magic on the basis of images from the ancient monument Shan Hai Jing is a tool for transforming into Human, as an omnipotent being.

2.14.7. "Superpowers" are really normal abilities of a human being; although, in potential. And, in this civilization, realization of this potential gets blocked.

2.14.8. Ancient myths, legends, fairy-tales were not thought-up at all.

2.14.9. Internal alchemy on the basis on the ancient Chinese monument Shan Hai Jing, which turned out to be not just an encyclopedia of people, but also a collection of magical recipes for fundamental transformation of *Homo sapiens*.

2.14.10. Connection of the ancient mystical traditions of the indigenous peoples of the Americas with the Catalog of Human Population.

2.14.11. Does the so-called "fate" of a human exist, and how to find it out without turning to numerous charlatans.

2.14.12. Chimeras: what they really are.

2.14.13. The world of spirits as reality.

PART 3

AHNENERBE—YOUR KILLER IS UNDER YOUR SKIN

The following text is taken from the book titled AHNENERBE—Your Killer Is Under Your Skin.

Sketch © 2000 Andrey Davydov

SECTION 1

Papier Mâché Apples

A human and a computer have a similarity: both do not work without software. A computer without software is a pile of useless metal. A human without software is a body, in which automatic physiological processes support life, but he cannot function, he is “a vegetable.” Today many people are familiar with computer software, but no one is familiar with “software” of a human. Although human software is actually the “soul” that throughout many centuries people are looking for so hard and cannot find. Software and psyche (soul) are synonyms. The role of human software is enormous: it is the basis of functioning of human psychophysiology.

This civilization has come up with a lot of modifications of artificial software without coming close to even approaching the question of every person having natural software, given by nature at the time of birth. People had to live somehow. Humans differ from animals: software of animals works automatically since the moment birth, it is not necessary to comprehend and study it, it just works. This is different with humans. So, artificial software was invented instead of natural. Although it differs from natural like papier-mâché or plastic apples differ from natural apples.

This civilization in fact forced humanity to eat plastic apples by creating artificial software. The outcome is easy to predict: such diet leads to imminent death. This is exactly what is happening. People of this civilization live for seventy to ninety years on average. Meanwhile physiologists state that the human body is designed by nature to live a healthy (and youthful!) life of several hundred of years. Instead, we have colossally rapid wear of psychophysiology, premature aging and death. Due to these "apples," inborn human potential is realized by one to two percent at best. Just imagine: everyday people eat only plastic apples. The result, however, is extended through time, and it is similar to the effect of a person taking daily micro dozes of poison.

What about the natural software? And, why no one knows about it? To know there needs to be a source of information. The source of knowledge about software of *Homo sapiens* as a textbook, from which each person can learn his or her natural program is not available in this civilization. Uncovering it without a source of information by means of "diving deep within oneself" does not work. Instead of knowledge there is only imagination. Without a "textbook" or, better said, instructions, a human simply cannot comprehend him or her self. The intellect of modern *Homo sapiens* does not work. He is blocked by artificial software. Completely! You do not believe this, are filled with outrage, protest, wondering what do people think with then? The answer is: with nothing because people do not think.

Instead of intellect people have ancestors' heritage, "Ahnenerbe." And, the word called "thought" is a file from this "Golden collection." People do not even have imagination. All imagination is the fruit of the "great inheritance," and that is all. So, getting to know his own self is not the only challenge that modern human intellect failed to achieve.

SECTION 2

"Frankenstein's Monsters"

Based on what software do you live if you are not familiar with your real self? Doctors and geneticists speak of heredity, psychologists and sociologists about the role of education and society in a person's life. Both are right: your software is the summed up heritage passed on to you by parents, educators, and culture. But, where are you? Unfortunately, you are nowhere.

Natural human software consists of natural images. Artificial software consists of artificial images. However, the foundation of both is an Image. Not without reason the great Lao Tzu wrote in the Tao Te Ching: "An Image is the basis of all things." And, the early Bible, likely until it was rewritten, instead of "In the beginning was the Word" stated: "In the beginning there was the Image." The Bible was translated from the Greek language, and in Greek philosophy 'eidos' (an image, a prototype, an archetype) does not occupy the last place at all.

An image is the basis of software, including software of a human. This has to do with the mechanisms of functioning of cells. Geneticists say that communication between cells takes place on the basis of video information exchange. Video information is images.

On what images does modern humanity live? There are a lot of them. Each era, every culture invents and "installs" various images in people. There were Apollo or Odin, then Zorro, then Batman, and so on. However, national (cultural) images get imposed on top of the initial basis, on that what is given to each person by those people, who raised and educated him. In most cases it is parents. However, it can be grandparents, or caregivers at an orphanage. The important thing here is that a person who raises a child gives him his own (upbringer's) philosophy as a recipe of life, as a lifestyle; recipes on what, when and how he should behave in different situations. For short, we called it "Ahnenerbe."

Also, it is important to keep in mind that natural software still works. One percent of it at best, but it works in every person. Life only on an "apple diet" is not possible in principle. Therefore, the natural program of a child and the natural program of a parent (upbringer) both work, only very weakly.

Therefore, modern people can be compared to the Frankenstein monster. This monster was also created from a huge number of different fragments of different human bodies. Only in the case of a person's software, it consists of a huge number of different images. Judge for yourself: parents' images, educators' images, a great number of images that get thrown into a person from culture, plus his natural images, of which he knows nothing.

This "Frankenstein" is a terrible mix for human cells. Only because of his enormous natural potential a human being can live on this monstrous mix for a few decades. However, he constantly gets sick, ages steadily, and often feels very unhappy. But, anyway, so far people of this civilization had nothing else except artificial Frankenstein software, so they had no choice. People had to believe that their artificial software is what they are.

It is phenomenal that in fact a person believes that artificially created and installed program is nothing other than him or her self.

Although logically, how can one consider mother's, father's, grandparents', aunts', uncles', upbringers', teachers', favorite characters from literature and movie altogether as his or her own self? People do this anyway. Parents' software is perceived as your own, one hundred percent. And, this happens regardless of the obvious fact that not you, but your parents and the civilization in general are the developers of your software.

SECTION 3

An Apple Does Not Fall Far From The Tree

By “parents” we are referring to those people, who took part in your upbringing. Regardless of who gave birth to you, it is far more important who raised you.

Nature made it so that unlike it is in the world of animals and plants, children of humans are not direct continuations of their parents. In human society, a hippo gives birth to an eagle, and fish to a horse.

Every child is born with a program, a natural one. It is recorded by natural images. However, unlike it is with animals, it is necessary to develop this natural program. In this, animals and humans are very different. The difference is that nature created animals to live automatically one hundred percent: they are born with a program and automatically live their whole lives because of it. For humans it is very different.

However, this civilization knows nothing about natural programs. Upbringers (parents) educate all children in exactly the same manner. Therefore, artificial software gets implanted in every human being from early childhood. And, because parents in this civilization are not familiar with natural programs of their children, instead of activating them, they "install" artificial software.

The purpose of this "install" is to give a child something to help him or her somehow function in this world. Artificial software thus replaces the natural program.

However, any software, whether natural or artificial, is software of a person, like a living computer. Therefore, a program must cover all six factors of his or her existence (intellectual, physical, emotional, sexual, nutritional, and environmental). Artificial software covers them all, while harmfully affecting the entire psychophysiology. Therefore, you accept it as the core, the essence of your individuality, and it defines your entire lifestyle.

What are the results of such "upbringing"? You can discover this from your own experience. Try to raise a newborn kitten like a fish, or like a chick. The result will be the same: a corpse. No kitten or chick will live long after being submerged in an aquarium. This is the result that each person of this civilization gets. Simply because a human is a being, who can still somehow survive after such damaging "upbringing." As we have shown, for animals such a level of violation in their upbringing would lead to death within a very short period of time. The killer under your skin (artificial software) kills very slowly. In humans, this process is extended through time, and so parents and children think that everything is fine, everything is good. Animals can only envy the powerful system of survival of human organism.

But still, the upbringing and education of a person that is not based on his/her natural program is always damaging to a greater or a lesser extent. In fact, the actions in this process cannot even be called education. The product, the "output" is not a Human, but a weakened, unhealthy being since childhood, an unwise creature, to put it mildly. This created a vicious circle, nonsense: one underdeveloped, damaged since childhood, anomalous creature attempts to educate another.

"Installation" of artificial program happens in the following way. During childhood, you perceived your parents as heavenly beings, as gods since satisfaction of all your needs (food, clothing, shelter, protection, education and so on) depended on them. And, that means that your life as such depended on them, hence the perception of parents as gods. After all, gods give life, and mom and dad were that to you.

And, they wasted no time. Their task was to teach you how to survive somehow in this world. However, they did not have the methods, did not have information on how to do it right. Hence, you inherited the most primitive (Neanderthal) philosophical base, and a simple tool called "a carrot and a stick."

SECTION 4

Listen To Mommy—Kill Yourself!

In adulthood a mother can be viewed differently, ranging from deepest respect and love to obvious hatred. The same goes for any other subject: an educator, a father, a brother, a grandmother, or a caretaker at an orphanage. "Love for mom" exists in any case as something immutable. Because it is love for particular sensations, not love for a particular person.

After all, what does a person value the most? A person values pleasant, comfortable feelings the most, regardless of who or what provides them.

Love for anyone, and especially for mother is based on feelings, sensations. A child needs the attention of adults. They are the ones who satisfy all of his needs until a certain age. If they do not, then a child simply dies. And, satisfaction of any needs provides a feeling of happiness. And, this is also a purely natural mechanism. That is why every satisfaction of every need of a child gets forever fixated in his mind, his memory, as nothing else, but happiness. And, this happiness is given to him by a certain specific person and people.

This childish feeling of complete happiness is the main (and the only!) reason for love for mother and her replacements. For a human, if there is a feeling of happiness, then there is bonding. By the way, all "engineers of human souls" and "psycho-technologists" know and successfully use this. Power over any person, first of all, is the power over his feelings.

A mother is the favorite and dearest being primarily because during childhood happiness ties to her. If there were no such feelings very firmly imprinted in psyche during childhood, then mom would be perceived on the same level as all others, and would not cause such high manifestations of affection.

For example, if some technical devices raised a child, satisfied his needs and caused happy feelings, then there would also be bonding. However, mothers have an advantage: they are alive. And, this means that a mother has emotions and transmits them to a child. (However, purposes of that may be different.) Every mother absolutely consciously "ties" her child to herself on the level of sensations, by caressing, hugging, saying some words. And, by doing that, sometimes absolutely sincerely, she naturally increases the value of her "product" for a child. No human being can stand against such transmissions, especially if they are done on a regular basis. And, when it comes to babies or toddlers ... there is no chance not to have the most tender feelings for mother for life.

A mother can be any and can perceive a child in any way, and a child can perceive her in any way. However, a human values feelings from childhood, and that is all. This means that regardless of his age, again and again he will try to relive them. A mother is just a person with whom it is easiest to remember these feelings. Hence, frequent communications with her, which often are actually selfish on her part.

However, regardless of being selfish or not, communications with mother are also very damaging; it is even possible to say—fatally destructive.

Since a child must pay a price for satisfaction of his needs, as well as for affection. Everyone is familiar with this: "If you don't do this, I won't give you that." Often, this is real blackmail because a disobedient child or a child, who does not meet the expectations of his mother, sometimes loses not just satisfaction of his whims as well as entertainment, but also satisfaction of his basic needs, and first of all, emotional ones. He needs a kind, gentle, loving relationship with his mother and right away gets a "bill": "Well then, do this" or "Be like this." If not, then the child hears: "You are bad! Mommy does not love you!" And, what could be bitterer?

However, more than just obedience lies behind the seemingly harmless requirements of his mother. These requirements can be, for example, to eat something, to behave in a certain way, often in spite of his natural movements, or to perform certain actions. In essence, this is a toolkit for a child's suicide. Since, what does a mother generally require from him? For one thing, she requires him to perform some civilizational requirements. She requires him to conform to her personal preferences, expectations, and thoughts on how something should be done. All this is strongly incompatible with the requirements of the psychophysiology of a child, with his personal natural characteristics and particularities. Of course, such conditions cannot be considered proper.

Figuratively speaking, mother-horse not just feeds vegetables to her child-tiger, but also requires that he loves them, and later on becomes a vegetarian like she is. Or, mother-penguin or mother-reindeer persistently tries to make her child-snake "stronger" by pouring cold water over him, threatening with punishment if he refuses to obey. In nature, the outcome is clear: death. However, do not rejoice too much: in the case of a human it is also death, only a slow one, stretched through time for years, with diseases, sufferings, and mental disorders along the way.

Each mother has her own natural program, among other things. And, in 99.9% out of 100% it is different from her child's program. However, a mother does not know about this. She thinks that the child is her continuation. She teaches him, as she believes, for his benefit: "Do as I do, live like I live." And, this is the main tragedy.

Left alone, without a mother, sometimes a person strives to follow what he was taught with even greater force. Why? Because he wants happiness with the same force, and regularly, but mom is far away. The only thing left to do in order to again "come back" to childhood is to live as mother told, to implement all of her teachings. And, for that, get a feeling of happiness. (However, not a real one, only as a record in memory, but pleasant nevertheless.)

And, what does it mean to follow mom's teachings? It means to kill yourself again and again. Consider several examples of analogues followings of parents' teachings: a tiger chewing on grass instead of eating meat; a horse diving into water and going down all the way to the bottom; a snake crawling away to live in ice; a bear basking in the sun in his winter coat all day at ninety degrees temperature (Fahrenheit). Suicide? Yes. And, not even for happiness, but for an illusion, a mirage.

A structure, which you consider your individuality, was created on this primitive basis. Although it is, in fact, just a set of reflexes "hammered" into you by "a carrot and a stick." The bottom line is that you think that everything that was "hammered" into you is your own, your "self." And, you are willing to stand by it to death, and you call it my "self", my personality, my individuality, and a bunch of other words that you yourself do not know the meaning of.

SECTION 5

"I Am Me!"

That what you consider to be your self, your real self, your own self—is not you.

However, despite all the facts and common sense, every person is convinced that all his actions and their consequences are fruits of his own decisions and choices. Even within yourself it is too painful and unpleasant to admit the fact that you are not the boss of yourself. A firm belief that you are not a replica of your parents and that you are an Individual occurs for one reason only. As you like to think, you have every reason to believe that you are your own creation. You say: "What parental philosophy? I forgot about it a long time ago!" That is correct, you forgot.

In addition, parents' "Frankenstein," which formed at five to ten years of age, certainly arranges himself "for himself" during the "rebellious" time of adolescence, and therefore adds his own fantasies to that what he wants to be, and nothing more. However, this is a minor correction to your already formed software. (Although it is very difficult to call it "your" because your parents and the civilization as a whole are the developers of this software, not you.)

Thus, after a variety of personal additions, the artificial program gets locked in. From that moment it begins to present a concrete construction. Parental "Golden Collection" becomes your way of life, and is perceived as your personal, individual path. And, this collection is not doubted, and does not go through cardinal revisions or changes.

Parents raise their children somehow (somehow because an individual recipe, different for each child is required for a different quality of upbringing.) At the age of eighteen to twenty parents say to their offspring: "That's it, you are an adult! You know everything!" You nod your head and go "into life" without even suspecting that your life is a life of a zombie, who will execute your parents' program implanted in you even to your own harm.

You were always convinced that you exist as an Individual, as a set of qualities, functionalities, and preferences. You called this (and continue to call it) "me." In fact, this is your personal "Frankenstein," which has been installed in you without asking your opinion.

And, in order to protect the installed program from possible changes, your desire to study has been cut off. Up until fifteen to twenty years of age a person studies somehow. However, after this age this process gets cut once you enter the category of adults, who know everything. All learning of life recipes stops: how to acquire new knowledge and skills; how to gather information, analyze it, test it out; how to find new variants, more suitable for present time, circumstances, situations; and so on. In adulthood, children often feel that the knowledge that was passed down from parents is not sufficient to live a full life, and sometimes even to simply survive in this world. They clearly feel that vital recipes "gifted" by parents do not work. However, they cannot do anything about it: "You're an adult!" And, the only thing that remains is to admit this.

Even in situations when there are opportunities to begin a new, better life by all indicators, people often ignore all the good things, acquisition of new skills and abilities, and prefer not to change the old way of life.

After all, this way of life provides such pleasant feelings!

SECTION 6

7.5 Billion Pavlov's Dogs

You do not know the secret of your tender affection towards, as you think, your own habits, opinions, preferences, and lifestyle. And, you have no desire to know anything about it. Any attempt to consider the mechanisms of your "Ahnenerbe," and even more so, any criticism of it or a suggestion to abandon it is perceived as a direct attack on your Individuality: the most precious, the most sacred, almost like life itself. And, this happens not at the level of the intellect, but at an even deeper level—on the level of sensations, feelings. (Therefore, it is impossible to register the source, to realize what is happening, let alone intervene.)

As for what are automatic responses in the form of sensations in response to a stimulus, one is familiar from textbooks on biology (experiments with Pavlov's dogs). This is the principle of your personal functioning, and the only one. For this reason, Pavlov's dog compared to you is an exemplary of consciousness.

You were not even the one, who installed the defense of your parents' heritage. Protection from any attack (outside and inside) on software is "embedded" in it. This makes your "virus" absolutely untouchable. And, the basis of immunity is sensations, feelings. You were trained since childhood. You are not aware of this and live only on "Ahnenerbe" reflexes. During your childhood you were taught what is good and what is bad. And, using the method of "a carrot and a stick" you were forced to have pleasant feelings when you did that what your parents needed, and vice versa.

Hence the craze of returning to childhood: "Oh, being a child was so good!" Although by psychiatric standards this is pathology; and, this is inability to get rid of "hammered in" parents' heritage. For as soon as you try to do something differently, you immediately feel very uncomfortable, sometimes even physical pain, and you instantly go back to following the artificial program. You want this, you want that, but mom taught not to do this. So, a human tries to live as his own nature tells him to, but then he gets "an electric shock." And, all attempts to understand his own self, as well as his self-development end here.

Therefore, it is not possible to speak of self-development, or even minimal fundamental changes in you. At this stage of development of your consciousness you are hopeless. Changes that are available to you are to repaint your nails into a different color, change your wardrobe, change your hairstyle, or, as they say, image. Maximal change is to move to another country, change your profession or family. No other changes, dear machines.

SECTION 7

Painful Then Pleasant, Painful Then Pleasant—And, You Are In A Stall.

Due to this mechanism, you are not present. You are simply a collection of reflexes. You react to any external and internal event (stimulus) with a set of conditioned responses. And, these reactions are not simply inadequate, not just unnatural, or anti-nature. They are not yours. However, since you do not exist as an Individual (as these conditioned reflexes are you) you are tightly enclosed in the "circle of Samsara." And, without a clear knowledge of how it works you are doomed.

The way it works is very simple. A stimulus (whether it is a headache (internal) or an act of another person (external) or a natural phenomenon (external)) causes pleasant or unpleasant sensations within you. Psychologists have already convinced everyone that all sensations are "good or bad." They also explained that the main principle of behavior should be to avoid everything that brings negative feelings and that one should "strive to be positive."

SECTION 8

Making The Same Mistake, And With Pleasure!

When a person drives a car without watching the road, more often than not he hits something. Normally it is considered that this should be avoided and people tend not to repeat this the second time, for example, after an accident. However, they do not do this with their habits, and, even more so with the way of life. It is considered sacred. And, therefore, this cannot be cheated on, should not be altered, must be defended, protected from abuse and one must stand his ground. At times, even in the face of certain death. This is recorded in "Ahnenerbe" as an artificial program of any person.

Besides, modern people do not think about the real reasons behind their actions, as well as their consequences. Artificial software dictates that you must not think, question your actions and decisions, and that the only criterion are feelings: "like/dislike, nice/not nice, I want/I do not want." And, it is recorded in the program not to pay attention to the fact that such criterion does not relate to any psychophysiological needs of a human being as a natural object or that actions on this basis hurt and actually kill. You are a zombie, and artificial software does not care whether you are healthy, wealthy, happy or not. It does not at all care if you will survive or not. The main requirement of "Ahnenerbe" is that you execute this program. Even at the cost of your life.

"Ahnenerbe" is a great tool to manipulate the work of your entire psychophysiology. It works perfectly, and in no way makes you even suspect that someone is controlling you.

Meanwhile, you are an absolute one hundred percent slave, on all six factors.

You do not know anything about the world and about your selves because you were ordered to believe. You do not know how to think, analyze, and all your "thoughts" are just a walk to the internal "cinema," where you will be told and shown how to live and what to be. And, you will have nothing left other than to do that, what your "inner voice" told you, and reinforced with corresponding sensations and feelings.

By the most primitive method—training—all of you were fooled. And, no one told you that reactions ("pleasant"/"unpleasant") do not come from your body, but come from your head. And, that what works in your head is not intellect, but a so-called "virtual room," which broadcasts how you need to live every minute. And, it guides you because you were told that this "inner voice" is your intellect.

SECTION 9

Brains... Hello?

Figuratively speaking: where is your intellect looking? Nowhere. It is disabled, blocked by artificial software.

Every "normal person" is used to talking to himself. You ask a question inside yourself and get a response, or just hear a voice. It is highly probable that none of you thought about whose voice it is.

Do you think it is your own? You are deeply mistaken.

That what you consider your own thoughts, ideas, opinions, desires, motivations, and so forth is just a "pop-up window" of your "Golden collection" in your brain. It was installed there on purpose, to make you into a human-swine, who sees nothing, hears nothing, is inadequate, and unfamiliar with what the real work of intellect, the thinking process really is. The "inner voice," constantly droning something in your ears, is not a voice. If it were just a "radio" that would be not so bad. However, it is not a radio, as it is a television. This "television" is in front of your eyes twenty-four hours a day because it broadcasts even when you are asleep. It spins an "Ahnenerbe" movie in front of your inner eye, telling you how to live, what to feel, what to do and what not to do.

Your "I need to think" means only one thing: you are now going to "fall deep within yourself," and there you will sort through the files of your "Ahnenerbe." This leads to inability to fit into reality in 99% out of 100 cases. Because in the artificial software it is recorded that if something is happening, then in order to fit in it is, supposedly, necessary for you to fall deep into yourself to find the appropriate file, and to act according to what is recorded there. However, the trouble is that the parents' "treasure" does not contain files for every occasion and the number of files is limited. There, in this software, it is also recorded that if one word matches the keyword describing the situation, then act as recorded in that file. As a result, the reality is generally not taken into account, and even more so its details and components. The situation does not get analyzed. First of all, you have nothing to analyze with. Information is necessary in order to analyze, but instead of gathering it a person "falls into himself," and at this moment all his sensors shut off automatically. He no longer sees, hears, smells, or feels. For a while he completely falls out of that what is happening around him. (This process is called: "I am thinking.") And, when he returns—he acts inappropriately.

First of all because the operating system inherited from parents contains files from very different times, often from a different country, those people, traditions, laws and so on. And, when this "treasure" was "installed" in you, you were five to ten, or maximum fifteen years old. So, with all this baggage you go and act—a child with tools made to address problems that existed twenty, thirty, fifty years ago; with a toolkit that rusted and rotted a long time ago, and does not work. The result is always the same: failure and damage to yourself, your well-being to one degree or another.

Moreover, you do not exist as an Individual. Sooner or later manifests such a fact as, for example, the presence of so-called "hereditary diseases." Or, for example, it is said about a person: "She repeated her mother's life." (This especially manifests in women.) And, this is logical. The consequences will be the same when one inherits a way of life as a philosophy, as software. And, it is the same with health. Certain habits lead to very specific health problems.

Few people know this, but any person pays a huge price for characteristics like “Just like her mother” or “Just like his father.” First of all, they pay with their health, and, in some cases, with their lives.

SECTION 10

Kaspersky Cries For You

Parental software has no relation to the natural "software" of a person. If we consider the parents' "installation" by analogy with computers: it is a virus. Everybody knows what a computer virus does. Computer virus hinders the work of a computer, blocks, does not allow the computer to function qualitatively, and, in the end, sooner or later, finally breaks it. The same thing happens with people.

In computers, viruses present constant errors. "Ahnenerbe" as a virus does the same. And, what is human error? This is a permanent inability to fit into a situation, and no matter what type of situation: internal state, health, communications, nature, social systems, and so on. The result of errors is always damage and injuries of varying severities. Relationships issues, health, financial, and psychological problems are all results of mistakes. A mistake means poor performance of a system, which in this case is a "human."

However, with computers it is much easier. It runs poorly? Do not worry, just use a certain program and remove all viruses. In severe cases, just replace the hard disk and the computer will be as good as new. And, at the moment it is not possible to do this with a human because modern medicine and psychology are still discussing the problem of a soul: "Does it exist or not?" And, both sciences believe that there is no soul. Therefore, there is nothing they can help with. Moreover, they are convinced that every person is a unique individual, and that adoption of parental installation is normal.

As a result, knowledge of natural software is unavailable, not to mention the idea that each person has it. The only thing that works is viral "Ahnenerbe," as software. Technologies on how to "clean a person from viruses" and activate his natural software are not available either. Meanwhile, in addition to daily damage from the "Ahnenerbe" virus, the working part of the natural software conflicts with artificial software on a daily basis throughout life. It is not only uncomfortable, but also dangerous to sit on two chairs at once. Thus, you tear yourself to pieces on the inside. Every day micro crack after micro crack gets added, and you do not even notice this. This civilization calls it the natural aging process. However, in reality, it is not natural.

SECTION 11

You Pay For Stupidity, Not For Pleasure

It is important to immediately note that artificial software was not designed so that you live well, so that you live as a healthy, wealthy, and powerful person. It has a different goal: to make you a slave, who can be exploited, used in many ways while he functions. Then, he gets disposed of, and, notably, by his own hands.

All control, management of you is under your skin. Rather, it is much deeper—in your psyche. There exists a recorded sensation, a feeling for every image. For example, that a mother is good, that you must love and obey her, even if she rudely breaks your psyche and body simply because she did not know your program. She damages, breaks, rapes you, but you continue to love her. What is the secret of such inadequacy? It is very simple! A mother, along with other "educators" reinforced damaging, raping, and breaking by pleasant sensations. For example: do as mommy says, and she will give you not only a delicious candy, buy also toys, take you to the movies or a circus, but more importantly, she will caress you, hug you tightly, pat you on the head, and you will feel happy...

Here is a case study: a programmed carnivore was raised by parents-vegetarians; later on, even when mother-father were no longer standing behind his back at dinner, he did not care that his body did not work properly due to lack of programmed nutrition. The person focused only on his feelings. He monitored only whether something was pleasant or unpleasant to him. And, while (as he was told to in childhood) he ate another batch of "tasty vegetables," instead of meat, which his body needs, he felt the same warm, good, bright, joyful feeling as he did during childhood. And, he ignored the fact that such meals gave him constipation, indigestion, stomach pain, weakness, and lack of energy. He got a set of "hammered in" experiences during his childhood with the label "this is happiness," and that satisfied him. And, even the fact that this was not true happiness did not bother him.

Management of a human is management of his feelings, and nothing more. Give any person exactly the same warm, bright, pleasant feelings, which mom and dad gave him, do this once, twice, ten times and the person is yours without a doubt, for sure. And, a person is willing to pay any price for this feeling, even his health and life itself. Hence, the management technique: put anything under this feeling and a person will perform; perform on a reflex, without thinking, like a dog that reacts to electric shocks and appetizing meat.

Based on this uncomplicated principle, that way of life, which your parents wanted you to follow, was "hammered" into you, so that they would be comfortable with you and your presence would not be too troublesome. And, your parents were taught this by society, which in turn requires new and new zombie, just like you, who will do just that what society needs.

So, briefly, here is how the artificial software based on artificial images works. You have every right to go on living with the hope of happiness, which you will never get. Because happiness is to really live, freely; happiness is to have your own thoughts, your own feelings, and to satisfy your own desires. However, you do not have them; there is only artificial software, "Ahnenerbe."

However, happiness as such exists. However, it is not comparable to your "happy fantasies," which are more like euthanasia. Happiness comes to a human as a strong and truly pleasant feeling only when he exists and realizes himself as an Individual; when he lives his own, not his mother's life. However, no matter how desirable such life is, slaves-reflexes cannot allow themselves to live like this.

To those, who do not like this and are sick of being in the position of slaves, we invite to overpower the strong internal opposition or a headache, and ponder the mechanism, which is your killer, and which, in fact, is under your skin. At least try to understand that parents trained you to accept this mechanism as "your self." That you were trained regularly, every day, for many years, just like it is done with some trained animals, which are made to live in a way that is far from requirements of nature. In your case, the matter is complicated by the fact that, unlike animals, you have no idea how you actually must live.

In this sense, it would probably be a good start to first find out your natural software and, to try to re-tune and tune the performance of your psychophysiology according to your natural images. It is possible to truly think, to use consciousness only as a result of a certain specific process aimed at development. And, the recipe of development must vary strongly for each person. The main tool must be studying. This option is long, but as we found out, it is the most optimal. After some time, as you move in this direction, the results will not take too long.

Note: This text does not specify the fact that images of artificial software were on purpose created by men (males), even though they are transmitted through women. You can research who the authors of education theory are (meaning, their gender), and get acquainted with the concept of patriarchy.

EPILOGUE

Sketch © 2000 Andrey Davydov

Over the last forty years, we got enough of both praises and criticism. Therefore, it does not make sense to puzzle over the question of why after making a scientific discovery of this level no one is in a hurry to get the Nobel Prize and why none of us is doing anything to get popularity, public recognition and worldwide fame. We simply consider it more sensible to invest our time and energy in those processes, which will give results that are more important, meaningful for us personally.

Moreover, many years of relentless pursuits to pose us as strange oddballs, astrologers, fringe scientists are both touching and striking. This civilizational trick is so archaic that all that is left is to wonder how kiddies are not bored playing the same game. By game we mean the standard civilizational approach to resolve a situation, when someone creates or discovers something that is needed and useful for all humanity; and, not unimportantly, when a lot of money can be made on its introduction into practical lives of people. During his life the author of this kind of invention or discovery gets persistently harassed, maltreated, put down, persecuted and his work gets totally rejected: "Joe and I do not need it!" And, once he is in the grave, all of a sudden, "out of the blue" people remember about this product, begin to praise it, etc. Then, everyone starts using this product. And, more importantly, someone, but already not the author of this product, starts making big money on this product. There are countless cases like this throughout history of this civilization: in science, in art, and in inventions. The scenario of this cheap show repeats from one century to the next. On the other hand, what else can be expected of Neanderthals?!

Therefore, we do not expect recognition, fame or honors from people of this civilization. In general, we prefer to do without the attention to our personas. We are quite tired from attention. If you want to use the Catalog—use it, and if you do not want to use it—do not use it. This is none of our business. Recognition is not required in order to do what we are doing. We find the position of authors of the Catalog of Human Population more appealing: having left the instructions to *Homo sapiens* in this civilization, they, as it appears, happily forgot about the Catalog, about humanity, and about this planet. Perhaps they had more exciting things to do than telling and proving things to someone. In this sense, we understand them very well.

We are not preventing anyone from living in any way that he or she wants. However, we have every right to live the way we want. We do not want to live the life of animals, especially since we have the only source in this civilization, which was created, most likely, not by people and contains information about how not to be an animal. However, this is our personal choice, our philosophy, our position in life. And, we are not suggesting to anyone to live as we do. Following basic ethics, through our books we are only giving a chance to those, who also would prefer not to be animals, but have not found a recipe on how to do it in reality.

Despite that our lives are constantly in danger due to the subject of our research, no matter how long they last, we prefer to spend them following the path to the "Human" status. To us, this life and this prospect seem more worthy than miserable existence, which people of this civilization call by the word "life." And, for us this is more interesting than to play games, which people of this civilization engage in so ecstatically.

Our aspiration to live as befits a human, as well as our personal knowledge and qualities obtained as a result of studying the Catalog of Mountains and Seas, can neither be killed nor taken by deceit or force, or bought, or unlawfully appropriated; neither from us nor from any one of all those people, who, like us, engaged and engage in the study of the heritage of our ancestors—Shan Hai Jing and other ancient monuments. These sources are that inviolability, imperishability, eternity through contact with which a person meets with eternity in his own self. And, in the end, becomes part of this Eternity.

BIBLIOGRAPHY

Vasilyev, V. P. (1866). *Analiz kitayskikh iyeroglifov (Sostavlen dlya rukovodstva studentov professorom S.-Peterburgskogo universiteta Vasilyevym)* [Analysis Of Chinese Hieroglyphs (Prepared By Professor Vasilyev Of University Of St. Petersburg To Guide Students)]. St. Petersburg.

Georgiyevskiy, S. M. (1892). *Mificheskiye vozzreniya i mify kitaytsev (s tablitsami kitayskikh iyeroglifov)* [Mythical Views And Myths Of The Chinese (With Tables Of Chinese Hieroglyphs)]. St. Petersburg: Tipografiya I. N. Skorokhodova.

Hall, M. P. (1928). *The Secret Teachings Of All Ages: An Encyclopedic Outline Of Masonic, Hermetic, Qabbalistic and Rosicrucian Symbolical Philosophy*. San Francisco: H.S. Crocker Company, Inc. Scanned November 2001, by Hare, J. Retrieved September 10, 2014, from http://www.framsteget.net/gratis/TheSecretTeachingsOfAllAges.pdf.

Yanshina, E. M. (Trans.). (1977). *Katalog Gor I Morej (Shan Hai Tszin)* [Catalog of Mountains and Seas (Shan Hai Jing)]. Moscow: Nauka.

Losev, A. F., Meletinskii, E. M., Riftin, B. L., Toporov, V. N., Ivanov, V. V., Averintsev, S. S., ... Afanasyeva, V. K. (1980). S. A. Tokarev (Ed.). *Mify narodov mira* [Myths of the World] (Vols. 1-2.). Moscow: Sovetskaya Entsiklopediya.

Sovetskiy Entsiklopedicheskiy Slovar' (SES) [Soviet Encyclopedic Dictionary]. (2nd ed.). (1982). Moscow: Sovetskaya Entsiklopediya.

Oshanin, I. M. (Ed.). (1983). *Bol'shoy kitaysko-russkiy slovar' (BKRS)* [Large Chinese-Russian Dictionary (BKRS)] (Vols. 1-4.). Moscow: Nauka.

Frazer, J. G. (1984). M. K. Ryklin (Trans.). *The Golden Bough: A Study In Magic And Religion*. Moscow: Politizdat. (Original work published 1890 under the title *The Golden Bough: A Study in Comparative Religion*.)

Pavlenko, N. A. (1987). *Istoriya pis'ma* [History of Writing]. Minsk: Vysheyshaya Shkola.

Starostin, S. A. (1989). *Rekonstruktsiya drevnekitayskoy fonologicheskoy sistemy* [Reconstruction Of The Ancient Chinese Phonological System.]. Moscow: Nauka.

Jung, C. G. (1991). A. M. Rutkevich (Ed.). *Arkhetip i simvol* [Archetype And Symbol]. Moscow: Renessans.

Shchutsky, J. K. (1993). A. I. Kobzev (Ed.). *Kitayskaya klassicheskaya "Kniga peremen"* [Chinese Classical Book of Changes]. (2nd ed.). Moscow: Vostochnaya Literatura.

Gariaev, P. P. (1994). *Volnovoy genom* [Wave Genome]. Moscow: Obshchestvennaya Polza. ISBN 9785856170053 ISBN 9785856171005 [Available at libraries http://www.worldcat.org/title/volnovoi-genom/.]

Lukyanov, A. E. (1994). *Nachalo drevnekitayskoy filosofii: "I tszin", "Dao de tszin", "Lun' yuy"* [The Beginning Of The Ancient Chinese Philosophy: I Ching, Tao Te Ching, Lun Yu]. Moscow: Radiks. ISBN 9785864630266. [Available at multiple libraries http://www.worldcat.org/title/nachalo-drevnekitaiskoi-filosofii-i-tszin-dao-de-tszin-lun-iui/.]

Davydov, A., & corr. of ITAR-TASS Fedoruk, V. (1998). Corr. of RAO L. Verbitskaya & Assoc. Prof. B. Sokolova (Eds.), *Pervyy Rossiyskiy Filosofskiy Kongress: Chelovek – Filosofiya – Gumanizm* [First Russian Philosophical Congress. Human Being – Philosophy – Humanism]: *Vol. 7 Philosophy and Human Problem: Is Shan Hai Jing The Original Catalog Of Psychophysiological Human Structure?* St. Petersburg: Saint Petersburg State University Publishing House, 7, 355-357. ISBN 9785288018947. Presentation at the First Russian Philosophical Congress: Human Being – Philosophy – Humanism (in Russian), St. Petersburg (1997). [Available at: University of Michigan http://mirlyn.lib.umich.edu/Record/003947324 B4231.R6751997 (subscription required), St. Petersburg's Central City Public Library Named After V. V. Mayakovsky http://www.pl.spb.ru/structure/zali/ ББК87Ч-391 (subscription required), and multiple other libraries http://www.worldcat.org/title/chelovek-filosofiia-gumanizm-pervyi-rossiiskii-filosofskii-kongress/]

Otyugov, A. A. (1998). Corr. of RAO L. Verbitskaya & Assoc. Prof. B. Sokolova (Eds.), *Pervyy Rossiyskiy Filosofskiy Kongress: Chelovek – Filosofiya – Gumanizm* [First Russian Philosophical Congress. Human Being – Philosophy – Humanism]: *Vol. 7 Philosophy and Human Problem: Psychological Nature Of Mythological Consciousness.* St. Petersburg: SPSU Publishing House. ISBN 9785288018947. Presentation at the First Russian Philosophical Congress: Human Being – Philosophy – Humanism (in Russian), St. Petersburg. Retrieved from http://mirlyn.lib.umich.edu/Record/003947324 B4231.R6751997 (subscription required). [Also available at other libraries http://www.worldcat.org/title/chelovek-filosofiia-gumanizm-pervyi-rossiiskii-filosofskii-kongress/.]

Torchinov, E. A. (1998). *Religii mira. Opyt zapredel'nogo (transpersonal'nyye sostoyaniya i psikhotekhnika).* [Religions of the World. Experience of the Transcendence (Transpersonal States and Psychotechnique)]. St. Petersburg: Tsentr "Peterburgskoye Vostokovedeniye". ISBN 9785858030782 [Available at libraries http://www.worldcat.org/title/religii-mira-opyt-zapredelnogo-transpersonalnye-sostojanija-i-psihotehnika/.]

Zarochentsev, K. D., & Khudyakov, A.I. (1998). Corr. of RAO L. Verbitskaya & Assoc. Prof. B. Sokolova (Eds.), *Pervyy Rossiyskiy Filosofskiy Kongress: Chelovek – Filosofiya – Gumanizm* [First Russian Philosophical Congress. Human Being – Philosophy – Humanism]: *Vol. 7 Philosophy and Human Problem: The Crisis Of Modern Psychodiagnostics*. St. Petersburg: SPSU Publishing House. ISBN 9785288018947. Presentation at the First Russian Philosophical Congress: Human Being – Philosophy – Humanism (in Russian), St. Petersburg. Retrieved from http://mirlyn.lib.umich.edu/Record/003947324 B4231.R6751997 (subscription required). [Also available at other libraries http://www.worldcat.org/title/chelovek-filosofiia-gumanizm-pervyi-rossiiskii-filosofskii-kongress/.]

Davydov, A., & Fedoruk, V. (1999). *Shan Khay Tszin: Mify Ili Struktura Psikhiki?* [Shan Hai Jing: Myths Or Structure Of Psyche?]. Moscow: *Power Of Spirit*, 32-35.

Davydov, A. (2002). *Katalog Chelovecheskoy Populyatsii* [*Catalog of Human Population*]. Presentation at the International Conference on Prospects of Preservation and Development of Unified Civilization of the Planet: Culture, Ecology, Cosmos (in Russian), Moscow (2002). [Mention in *Parlamentskaya Gazeta*: "...The most luminous were reports by ... A. Davydov about the program that he developed, which opens up the best, but often hidden qualities of a human..." (Umorov, I. (2002). *Chelovechestvu Otpushcheno Tol'ko Tri Goda* [Humanity Has Only Three Years]. Moscow: *Parlamentskaya Gazeta, Sphere of Reason*, 129 (1008), para. 4. Retrieved 13 May 2015 from http://old.pnp.ru/archive/10082921.html.)]

Ezhov, V. V. (2003). *Mify Drevnego Kitaya* [Myths of Ancient China]. (Ill. *Myths of Nations of the World*) Moscow: Astrel.

Wilhelm, R., & Wilhelm H. (2003). V.B. Kurnosova (Trans.). *Ponimaniye «I tszin». Antologiya*. [Understanding The I Ching. Anthology.]. (2nd ed.). *Tradition, Religion, Culture*. Moscow: Aleteya.

Terentev-Katansky, A. (2004). *Illyustratsii k kitayskomu bestiariyu: Mifologicheskiye zhivotnyye drevnego Kitaya* [Illustrations to Chinese Bestiary: Mythological Animals of Ancient China]. St. Petersburg: Forma T.

Nikitina, T. N., & Zaytsev, V.P. (Eds.). (2009). *Slovar' drevnekitayskikh iyeroglifov = 古代漢語字典: s prilozheniyem slovarya naiboleye chastotnykh omografov, vstrechayushchikhsya v drevnekitayskom tekste, sost. Ye. G. Ivanovoy* [The Dictionary Of Ancient Chinese Hieroglyphs = 古代 漢語 字典: With The Appended Dictionary Of Most Frequent Homographs Found In Ancient Chinese Text, Compiled By E. G. Ivanova]. St. Petersburg: Karo. ISBN 9785992504293

Davydov, A., & Skorbatyuk, O. (2014). K. Bazilevsky (Trans.). *AHNENERBE—Your Killer Is Under Your Skin* (Composed 2014. Original work published 2014 in Russian, ISBN 9781311356741.). San Diego, CA: HPA Press. ISBN 9781311266682

Davydov, A., & Skorbatyuk, O. (2014). K. Bazilevsky (Trans.). *IDEOLOGY OF RELIGIONS. Scientific Proof Of Existence Of "God": The Catalog Of Human Population.* (Original work published 2014 in Russian, ISBN 9781311946690.). San Diego, CA: HPA Press. ISBN 9781311413932 ISBN 9780988648593

OUR OTHER BOOKS RELATED TO OUR SCIENTIFIC RESEARCH

Monographic Series

Archetypal Pattern. Fundamentals of Non-Traditional Psychoanalysis.

Davydov, A., & Skorbatyuk, O. (2014). K. Bazilevsky (Ed.). Anonymous (Trans.). *Archetypal Pattern. Fundamentals of Non-Traditional Psychoanalysis*: *Vol. 1. From Carl Gustav Jung's Archetypes of the Collective Unconscious to Individual Archetypal Patterns.* (Composed 2005. Original work published 2013 in Russian, ISBN 9781301447688.). San Diego, CA: HPA Press. ISBN 9781311820082

Davydov, A., & Skorbatyuk, O. (2014). K. Bazilevsky (Trans.). *Archetypal Pattern. Fundamentals of Non-Traditional Psychoanalysis*: *Vol. 2. Can Archetypal Images Contain Chimeras?* (Composed 2005. Original work published 2013 in Russian, ISBN 978130184859.). San Diego, CA: HPA Press. ISBN 9781310658570

Davydov, A., & Skorbatyuk, O. (2014). Arkhetipicheskiy Pattern. Osnovy Netraditsionnogo Psikhoanaliza [Archetypal Pattern. Fundamentals of Non-Traditional Psychoanalysis]: *Vol. 3. Archetype Semantics: How This Corresponds to the Concept of an 'Image'. How Archetypal Are Images?* (Composed 2005.). Marina Del Rey, CA: Catalog Of Human Souls GP. ISBN 9781301337309

Davydov, A., & Skorbatyuk, O. (2014). K. Bazilevsky (Trans.). *Archetypal Pattern. Fundamentals of Non-Traditional Psychoanalysis*: *Vol. 4. Society As A Community Of Manipulators And Their Subjects.* (Composed 2005. Original work published 2013 in Russian, ISBN 9781301399901.). San Diego, CA: HPA Press. ISBN 9781311809353

Catalog of Human Population - Non-Fiction Series

Individual (Subtype) Human Programs

Davydov, A., & Skorbatyuk, O. (2013). *Katalog Chelovecheskikh Dush: Programmnoye Obespecheniye Dushi Muzhchin/Zhenshchin, Rodivshikhsya <Data>* [Catalog of Human Souls: Software of Soul of Men/Women Born On <Date>] (Vols. 1-218. In Russian. Composed 2005-

2013.). Marina Del Rey, CA: Catalog Of Human Souls GP. [Available at http://www.humanpopulationacademy.org/pricing/ in all languages].

Human Manipulation Modes

Davydov, A., & Skorbatyuk, O. (2013-2014). *Katalog Chelovecheskikh Dush: Kak Podchinit' Muzhchin/Zhenshchin, Rozhdonnykh <Data>. Zhenskiy/Muzhskoy Manipulyativnyy Ctsenariy.* [Catalog of Human Souls: How To Subdue Men/Women Born On <Date>. Female/Male Manipulation Scenario.] (Vols. 1-39. In Russian. Composed 2005-2013.). Marina Del Rey, CA: Catalog Of Human Souls GP. [Available at http://www.humanpopulationacademy.org/pricing/ in all languages].

Ideologies

Davydov, A. (2014). K. Bazilevsky (Trans.). *Terrorism: A Concept For The ATC (The Commonwealth Of Independent States Anti-Terrorism Center).* (Composed 2001. Original work published 2014 in Russian, ISBN 9781311277848.). San Diego, CA: HPA Press. ISBN 9781310032189

Davydov, A. (2014). K. Bazilevsky (Trans.). *Ideology Of Monarchy. For Office Of The Head Of The Russian Imperial House, Her Imperial Highness Grand Duchess Maria Vladimirovna.* (Composed 2003. Original work published 2014 in Russian, ISBN 9781310150340.). San Diego, CA: HPA Press. ISBN 9781311970152

Davydov, A., & Skorbatyuk, O. (2014). K. Bazilevsky (Trans.). *Ideology Of Religions. Scientific Proof Of Existence Of "God": The Catalog Of Human Population.* (Original work published 2014 in Russian, ISBN 9781311946690.). San Diego, CA: HPA Press. ISBN 9781311413932 ISBN 9780988648593

Political Science

Davydov, A. (2014). K. Bazilevsky (Trans.). *Essence Of Political Ideologies And Their Role In The Historical Process (Political History Of Russia).* (Composed 2003. Original work published 2014 in Russian, ISBN 9781310199929.). San Diego, CA: HPA Press. ISBN 9781310199929

Davydov, A. (2014). K. Bazilevsky (Trans.). *Influence Of Psychophysiological Specifics Of A Leader On The Style Of Political Decision-Making.* (Composed 2003. Original work published 2014 in Russian, ISBN 9781310037832). San Diego, CA: HPA Press. ISBN 9781310104558

Davydov, A. (2014). K. Bazilevsky (Trans.). *Elitist Political Concepts.* (Composed 2005. Original work published 2014 in Russian, ISBN 9781310223228). San Diego, CA: HPA Press. ISBN 9781310822858

General Non-Fiction

Bazilevsky, K. (2012). *Human Population Academy: Laws of Human Nature Based on Shan Hai Jing Research Discoveries by A. Davydov & O. Skorbatyuk.* San Diego, CA: HPA Press. ISBN 9781301986781 ISBN 9780988648500

Davydov, A. (2013). *Shan Khay Tszin: Mify Ili Struktura Psikhiki?* [Shan Hai Jing: Myths Or Structure Of Psyche?] (Composed 1999. Originally pub. 1999 in Russian in Moscow: *Power Of Spirit*, 32-35.). Marina Del Rey, CA: Catalog Of Human Souls GP. ISBN 9781301590391

Davydov, A. (2013). *"Shan Khay Tszin" i "I Tszin" – Karta Psikhofiziologicheskoy Struktury Cheloveka?* [Shan Hai Jing and I Ching – Map of Human Psychophysiological Structure?] (Composed 2002.). Marina Del Rey, CA: Catalog Of Human Souls GP. ISBN 9781301510009

Davydov, A., & Skorbatyuk, O. (2014). K. Bazilevsky (Trans.). *AHNENERBE: Your Killer Is Under Your Skin* (Original work published 2014 in Russian, ISBN 9781311356741.). San Diego, CA: HPA Press. ISBN 9781311266682

A Man And A Woman – Non-Fiction Series

A Log With Legs Spread Wide

Davydov, A., & Skorbatyuk, O. (2014). K. Bazilevsky (Trans.). *A Log With Legs Spread Wide: Vol. 1. How Men Turn Women Into Nothing.* (Original work published 2014 in Russian, ISBN 9781310388125.). San Diego, CA: HPA Press. ISBN 9781311155771

Davydov, A., & Skorbatyuk, O. (2014). K. Bazilevsky (Trans.). *A Log With Legs Spread Wide: Vol. 2. How Goddesses Are Turned Into Logs. World History Of Turning Women Into Mats.* (Original work published 2014 in Russian, ISBN 9781311238894.). San Diego, CA: HPA Press. ISBN 9781311915603

Davydov, A., & Skorbatyuk, O. (2013). *A Log With Legs Spread Wide: Vol. 3. Women's Thirst For Power Over Men Is The Pathway To Become A*

Garbage. (Original work published 2013 in Russian, ISBN 9781301553075.). Marina Del Rey, CA: Catalog Of Human Souls GP. ISBN 9781301435500

Davydov, A., & Skorbatyuk, O. (2013). *A Log With Legs Spread Wide: Vol. 4. The Head – In The Underpants*. (Original work published 2013 in Russian, ISBN 9781301051281.). Marina Del Rey, CA: Catalog Of Human Souls GP.

Manipulative Games For Women

Davydov, A., & Skorbatyuk, O. (2013). *Manipulyativnyye Igry Dlya Zhenshchin* [Manipulative Games For Women]: *Vol. 1. March 23: Instruction for Exploitation of Men* (2nd ed., in Russian. Original work published 2005, Moscow: SNIALTotems. ISBN 9785716101333). Marina Del Rey, CA: Catalog Of Human Souls GP. ISBN 9781301803521

Davydov, A., & Skorbatyuk, O. (2013). *Manipulyativnyye Igry Dlya Zhenshchin* [Manipulative Games For Women]: *Vol. 2. April 6: Instruction for Exploitation of Men* (2nd ed., in Russian. Original work published 2005, Moscow: SNIALTotems. ISBN 9785716101302). Marina Del Rey, CA: Catalog Of Human Souls GP. ISBN 9781301069286

Davydov, A., & Skorbatyuk, O. (2013). *Manipulyativnyye Igry Dlya Zhenshchin* [Manipulative Games For Women]: *Vol. 3. October 13: Instruction for Exploitation of Men* (2nd ed., in Russian. Original work published 2005, Moscow: SNIALTotems. ISBN 9785716101326). Marina Del Rey, CA: Catalog Of Human Souls GP. ISBN 9781301900824

Davydov, A., & Skorbatyuk, O. (2013). *Manipulyativnyye Igry Dlya Zhenshchin* [Manipulative Games For Women]: *Vol. 4. December 7: Instruction for Exploitation of Men* (2nd ed., in Russian. Original work published 2005, Moscow: SNIALTotems. ISBN 9785716101319). Marina Del Rey, CA: Catalog Of Human Souls GP. ISBN 9781301413065

Secret Sexual Desires

Bazilevsky, K. (2013). *How To Seduce Men/Women Born On <Date> Or Secret Sexual Desires of 10 Million People: Demo From Shan Hai Jing Research Discoveries by A. Davydov & O. Skorbatyuk*. (Vols. 1-10). San Diego, CA: HPA Press.

Bazilevsky, K. (2013). *How To Seduce Men & Women Born On March 5 Or Secret Sexual Desires of 20 Million People: Demo From Shan Hai Jing Research Discoveries by A. Davydov & O. Skorbatyuk*. San Diego, CA: HPA Press. ISBN 9781301087204

Bazilevsky, K. (2013). *Secret Sexual Desires of 100 Million People—Seduction Recipes For Men & Women: Demos From Shan Hai Jing Research Discoveries by A. Davydov & O. Skorbatyuk*. San Diego, CA: HPA Press. ISBN 9780988648579 ISBN 9781301135035 ISBN 9780988648586

A list of other publications related to our scientific research can be found at http://www.humanpopulationacademy.org/publications/.

CONNECT WITH US

1. Visit our official website.

Human Population Academy and Special Scientific Info-Analytical Laboratory—Catalog of Human Souls: https://www.HumanPopulationAcademy.org

2. Connect with us on social networks.

- *Facebook* - http://www.facebook.com/HumanPopulationAcademy (Note: you must be logged in to *Facebook* in order to access this page.)
- *YouTube* - http://www.youtube.com/user/HumanPopulAcademy
- *Google+* - http://plus.google.com/+HumanpopulationacademyOrghumannature
- *LinkedIn* - http://www.linkedin.com/company/2484433
- *Pinterest* - http://pinterest.com/humanpopacademy/
- *Twitter* - http://twitter.com/HumanPopAcademy

3. Contact us.

You can find out how to contact us at the Human Population Academy's website under Contacts (see http://www.humanpopulationacademy.org/breakthrough-discovery/contacts/).

ABOUT US

Special Scientific Info-Analytical Laboratory—Catalog of Human Souls was founded by Andrey Davydov. The laboratory is engaged in research and decryption of the ancient Chinese monument Shan Hai Jing, as well as other ancient texts, and creation of the *Catalog of Human Population*. The technology of uncovering individual structures of psyche of *Homo sapiens* for this Catalog was developed by Andrey Davydov; it is not based on any existing domestic or foreign research, methods or theoretical concepts. The laboratory is a partner with the Human Population Academy.

Human Population Academy was founded by Kate Bazilevsky. The Academy's mission is to inform all of over 7 billion humans living on Earth about the discovery of the *Catalog of Human Population*. The Academy educates about the *Catalog of Human Population* (*Catalog of Human Souls*) and provides access to informational materials from this Catalog to the public.

LEADERSHIP

ANDREY DAVYDOV

Research Supervisor of the Special Scientific Info-Analytical Laboratory—Catalog of Human Souls

Andrey Davydov is an expert in Chinese culture, researcher of ancient texts, the author of scientific discovery of the *Catalog of Human Population* and the technology of decryption of the ancient Chinese monument Shan Hai

Jing as the *Catalog of Human Population*. He authored over 300 published books, including scientific monographs and ideologies. In 2012, he was granted political asylum in the USA due to persecution by a group of employees of the Federal Security Service of Russian Federation (FSB, formerly KGB), who decided to expropriate his research product—the *Catalog of Human Population*.

OLGA SKORBATYUK

Senior Analyst at the Special Scientific Info-Analytical Laboratory—Catalog of Human Souls

Olga Skorbatyuk is a professional psychologist, one of the developers of the *Catalog of Human Population*, the founder of Non-Traditional Psychoanalysis, and co-author of over 300 books and scientific articles. She was granted political asylum in the USA together with A. Davydov.

KATE BAZILEVSKY

Founder of the Human Population Academy, Junior Analyst at the Special Scientific Info-Analytical Laboratory—Catalog of Human Souls

Kate Bazilevsky is the director of the Human Population Academy, a Junior Analyst at the Catalog of Human Souls laboratory, an author and a translator of books about the *Catalog of Human Population*. She holds a degree in MIS and psychology. She founded the Human Population Academy in 2011 and a publishing company called HPA Press in 2012.

www.ingramcontent.com/pod-product-compliance
Lightning Source LLC
LaVergne TN
LVHW052314200926
848481LV00002B/557

* 9 7 8 0 9 9 6 7 3 1 2 0 1 *